PRAISE F[...]

How to Teach Quantum Physics to Your Dog

"Don't let Orzel's laid-back nature or clever sense of humor fool you—he is explaining some pretty serious stuff. A levelheaded and confident guide, he takes Emmy (and the reader) through everything from wave-particle duality and superpositions to quantum tunneling and the so-called 'many worlds' interpretation ('many worlds, many treats')."

—NewScientist.com

"It's hard to imagine a better way for the mathematically and scientifically challenged, in particular, to grasp basic quantum physics."

—*Booklist*

"This charming little book is a lighthearted and amusing way for laypeople to learn about one of the strangest and most important aspects of modern science. It is also a great resource for practicing 'quantum mechanics' who want new ideas on how to more effectively explain their work to the public."

—William D. Phillips, 1997 Nobel Laureate in Physics

"Professor Orzel has a gift for funny dialogue and straightforward explanation. In addition to the entertaining conversations with Emmy, there are fascinating explanations of how the theories behind quantum mechanics were developed and how a few have been tested."

—DogSpelledForward.com

"Dogs make the perfect sounding board for physics talk. . . . [Orzel's] cheerful discussion [is] a real treat."

—*Publishers Weekly*

"I've long believed that everyone should be familiar with the wonders of quantum mechanics. I had no idea that 'everyone' would include dogs! Chad Orzel's book is a fast-moving and fun introduction to some of the deepest mysteries of modern physics. And Emmy is a star."

—Sean Carroll, author of *From Eternity to Here*

"Chad Orzel teases out the mysterious and seemingly incomprehensible side of advanced physics and makes it comprehensible via one-sided monologues to even the most distractible: dogs, humans, and in my case even disdainful felines or somewhat puzzled infants."

—Tobias S. Buckell, author of *HALO: The Cole Protocol*

"An absolutely delightful book on many axes: first, its subject matter, quantum physics, is arguably the most mind-bending scientific subject we have; second, the device of the book . . . finally, third, it is extremely well-written, combining a scientist's rigor and accuracy with a natural raconteur's storytelling skill."

—boingboing.net

"Orzel's whimsical take on quantum physics is a delight, and Emmy is the perfect Everyman, posing the questions we'd all like to ask about the intricacies of this most esoteric of subjects."

—Jennifer Ouellette, author of *The Physics of the Buffyverse*

"Quantum physics is perhaps the most interesting and slipperiest scientific subject; who knew that Socratic discussion with an adorable dog was the key to unraveling it?"

—Cory Doctorow, author of *Little Brother* and coeditor of Boing Boing

"My dog Kodi tells me that Chad Orzel explains physics with far more clarity and humor than I ever did, and that now she's just keeping me around for my opposable thumbs. Thanks a lot, Chad."

—John Scalzi, author of *Old Man's War* and *The Rough Guide to the Universe*

HOW TO TEACH QUANTUM PHYSICS TO YOUR DOG

CHAD ORZEL

SCRIBNER

New York London Toronto Sydney New Delhi

SCRIBNER
A Division of Simon & Schuster, Inc.
1230 Avenue of the Americas
New York, NY 10020

First Scribner trade paperback edition December 2010

DESIGNED BY ERICH HOBBING

Library of Congress Control Number: 2009021073

ISBN 978-1-4165-7228-2
ISBN 978-1-4165-7229-9 (pbk)
ISBN 978-1-4165-7901-4 (ebook)

To Kate,
whose laugh
started the whole thing

Contents

HOW TO TEACH QUANTUM PHYSICS TO YOUR DOG

INTRODUCTION

Why Talk to Your Dog about Physics? An Introduction to Quantum Physics

The Mohawk-Hudson Humane Society has set up a little path through the woods near their facility outside Troy, so you can take a walk with a dog you're thinking of adopting. There's a bench on the side of the path in a small clearing, and I sit down to look at the dog I've taken out.

She sits down next to the bench, and pokes my hand with her nose, so I scratch behind her ears. My wife and I have looked at a bunch of dogs together, but Kate had to work, so I've been dispatched to pick out a dog by myself. This one seems like a good fit.

She's a year-old mixed-breed dog, German shepherd and something else. She's got the classic shepherd black and tan coloring, but she's small for a shepherd, and has floppy ears. The tag on her kennel door gave her name as "Princess," but that doesn't seem appropriate.

"What do you think, girl?" I ask. "What should we call you?"

"Call me Emmy!" she says.

"Why's that?"

"Because it's my name, silly."

Being called "silly" by a dog is a little surprising, but I guess she has a point. "Okay, I can't argue with that. So, do you want to come live with us?"

"Well, that depends," she says. "What's the critter situation like?"

"Beg pardon?"

"I like to chase things. Will there be critters for me to chase?"

"Well, yeah. We've got a good-sized yard, and there are lots of birds and squirrels, and the occasional rabbit."

"Ooooh! I like bunnies!" She wags her tail happily. "How about walks? Will I get walks?"

"Of course."

"And treats? I like treats."

"You'll get treats if you're a good dog."

She looks faintly offended. "I am a *very* good dog. You *will* give me treats. What do you do for a living?"

"What? Who's evaluating who, here?"

"I need to know if you deserve a dog as good as me." The name "Princess" may have been more apt than I thought. "What do you do for a living?"

"Well, my wife, Kate, is a lawyer, and I'm a professor of physics at Union College, over in Schenectady. I teach and do research in atomic physics and quantum optics."

"Quantum what?"

"Quantum optics. Broadly defined, it's the study of the interaction between light and atoms in situations where you have to describe one or both of them using quantum physics."

"That sounds complicated."

"It is, but it's fascinating stuff. Quantum physics has all sorts of weird and wonderful properties. Particles behave like waves, and waves behave like particles. Particle properties are indeterminate until you measure them. Empty space is full of 'virtual particles' popping in and out of existence. It's really cool."

"Hmmm." She looks thoughtful, then says, "One last test."

"What's that?"

"Rub my belly." She flops over on her back, and I reach down to rub her belly. After a minute of that, she stands up, shakes herself off, and says "Okay, you're pretty good. Let's go home."

We head back to the kennel to fill out the adoption paperwork. As we're walking, she says, "Quantum physics, huh? I'll have to learn something about that."

"Well, I'd be happy to explain it to you sometime."

Like most dog owners, I spend a lot of time talking to my dog. Most of our conversations are fairly typical—don't eat that, don't climb on the furniture, let's go for a walk. Some of our conversations, though, are about quantum physics.

Why do I talk to my dog about quantum physics? Well, it's what I do for a living: I'm a college physics professor. As a result, I spend a lot of time thinking about quantum physics.

What is quantum physics? Quantum physics is one part of "modern physics," meaning physics based on laws discovered after about 1900. Laws and principles of physics that were developed before about 1900 are considered "classical" physics.

Classical physics is the physics of everyday objects—tennis balls and squeaky toys, stoves and ice cubes, magnets and electrical wiring. Classical laws of motion govern the motion of anything large enough to see with the naked eye. Classical thermodynamics explains the physics of heating and cooling objects, and the operation of engines and refrigerators. Classical electromagnetism explains the behavior of lightbulbs, radios, and magnets.

Modern physics describes the stranger world that we see when we go beyond the everyday. This world was first revealed in experiments done in the late 1800s and early 1900s, which cannot be explained with classical laws of physics. New fields with different rules needed to be developed.

Modern physics is divided into two parts, each representing a radical departure from classical rules. One part, **relativity,** deals with objects that move very fast, or are in the presence of strong gravitational forces. Albert Einstein introduced relativity in 1905, and it's a fascinating subject in its own right, but beyond the scope of this book.

The other part of modern physics is what I talk to my dog about. **Quantum physics** or **quantum mechanics*** is the name given to the part of modern physics dealing with light and things that are very small—molecules, single atoms, subatomic particles. Max Planck coined the word "quantum" in 1900, and Einstein won the Nobel Prize for presenting the first quantum theory of light.† The full theory of quantum mechanics was developed over the next thirty years or so.

The people who made the theory, from early pioneers like Planck and Niels Bohr, who made the first quantum model of the hydrogen atom, to later visionaries like Richard Feynman and Julian Schwinger, who each independently worked out what we now call "quantum electrodynamics" (QED), are rightly regarded as titans of physics. Some elements of quantum theory have even escaped the realm of physics and captured the popular imagination, like Werner Heisenberg's uncertainty principle, Erwin Schrödinger's cat paradox, and the parallel universes of Hugh Everett's many-worlds interpretation.

Modern life would be impossible without quantum mechanics. Without an understanding of the quantum nature of the electron, it would be impossible to make the semiconductor chips that run our computers. Without an understanding of the quantum nature of light and atoms, it would be impossible to make the lasers we use to send messages over fiber-optic communication lines.

Quantum theory's effect on science goes beyond the merely

*The terms "quantum physics," "quantum theory," and "quantum mechanics" are more or less interchangeable.

†Inventing relativity didn't exactly hurt, but the official reason for Einstein's Nobel was his quantum theory of the photoelectric effect (page 22).

practical—it forces physicists to grapple with issues of philosophy. Quantum physics places limits on what we can know about the universe and the properties of objects in it. Quantum mechanics even changes our understanding of what it means to make a measurement. It requires a complete rethinking of the nature of reality at the most fundamental level.

Quantum mechanics describes an utterly bizarre world, where nothing is certain and objects don't have definite properties until you measure them. It's a world where distant objects are connected in strange ways, where there are entire universes with different histories right next to our own, and where "virtual particles" pop in and out of existence in otherwise empty space.

Quantum physics may sound like the stuff of fantasy fiction, but it's science. The world described in quantum theory is our world, at a microscopic scale.* The strange effects predicted by quantum physics are real, with real consequences and applications. Quantum theory has been tested to an incredible level of precision, making it the most accurately tested theory in the history of scientific theories. Even its strangest predictions have been verified experimentally (as we'll see in chapters 7, 8, and 9).

So, quantum physics is neat stuff. But what does it have to do with dogs?

Dogs come to quantum physics in a better position than most humans. They approach the world with fewer preconceptions than humans, and always expect the unexpected. A dog can walk down the same street every day for a year, and it will be a new experience every day. Every rock, every bush, every tree will be sniffed as if it had never been sniffed before.

If dog treats appeared out of empty space in the middle of

*"Microscopic" for a physicist means anything too small to be seen with the naked eye. This covers a range from bacteria to atoms to electrons. It's a wide range of sizes, but physicists think it would be confusing to have more than one word for small things.

a kitchen, a human would freak out, but a dog would take it in stride. Indeed, for most dogs, the spontaneous generation of treats would be vindication—they always expect treats to appear at any moment, for no obvious reason.

Quantum mechanics seems baffling and troubling to humans because it confounds our commonsense expectations about how the world works. Dogs are a much more receptive audience. The everyday world is a strange and marvelous place to a dog, and the predictions of quantum theory are no stranger or more marvelous than, say, the operation of a doorknob.*

Discussing quantum physics with my dog is useful because it helps me see how to discuss quantum mechanics with humans. Part of learning quantum mechanics is learning to think like a dog. If you can look at the world the way a dog does, as an endless source of surprise and wonder, then quantum mechanics will seem a lot more approachable.

This book reproduces a series of conversations with my dog about quantum physics. Each conversation is followed by a detailed discussion of the physics involved, aimed at interested human readers. The topics range from ideas many people have heard of, like particle-wave duality (chapter 1) and the uncertainty principle (chapter 2), to the more advanced ideas of virtual particles and QED (chapter 9). These explanations include discussion of both the weird predictions of the theory (both practical and philosophical), and the experiments that demonstrate these predictions. They're selected for what dogs find most interesting and also illustrate the parts that humans find surprising.

"I don't know. I think it needs . . . more."

"More what?"

*Which unquestionably follows classical rules, but does, alas, require opposable thumbs to operate.

"More me. You don't talk about the fact that I'm an exceptionally smart dog."

"Well, okay—"

"And exceptionally cute, too."

"Sure, but—"

"And don't forget good. I'm way better than those other dogs."

"What other dogs?"

"Dogs who aren't me."

"Look, this is really a book about physics, not a book about you."

"Well, it ought to be more about me, that's all I'm saying."

"It's not, and you'll just have to live with that."

"Okay, fine. You need my help with the physics stuff, though."

"What do you mean?"

"Well, sometimes you leave some stuff out, and don't answer all of my questions. You shouldn't do that."

"Like what? Give me an example."

"Ummm . . . I can't think of one now. If you read it to me, though, I'll point them out, and help fix them."

"Okay, that sounds fair. Here's what we'll do. We'll go over the book together, and if there are places where you think I've left stuff out, we can talk about them, and I'll put your comments in the book."

"Talk about them like we're doing now?"

"Yeah, like we're doing now."

"And you'll put the conversation in the book?"

"Yes, I will."

"In that case, we should talk about how I'm the very best, and I'm cute, and I should get more treats, and—"

"Okay, that's about enough of that."

"For now."

CHAPTER 1

Which Way? Both Ways: Particle-Wave Duality

We're out for a walk, when the dog spots a squirrel up ahead and takes off in pursuit. The squirrel flees into a yard and dodges around a small ornamental maple. Emmy doesn't alter her course in the slightest, and just before she slams into the tree, I pull her up short.

"What'd you do that for?" she asks, indignantly.

"What do you mean? You were about to run into a tree, and I stopped you."

"No I wasn't." She looks off after the squirrel, now safely up a bigger tree on the other side of the yard. "Because of quantum."

We start walking again. "Okay, you're going to have to explain that," I say.

"Well, I have this plan," she says. "You know how when I chase the bunnies in the backyard, when I run to the right of the pond, they go left, and get away?"

"Yes."

"And when I run to the left of the pond, they go right, and get away?"

"Yes."

"Well, I've thought of a new way to run, so they can't escape."

"What, through the middle of the pond?" It's only about eight inches deep and a couple of feet across.

"No, silly. I'm going to go both ways. I'll trap the bunnies between me."

"Uh-huh. That's an . . . interesting theory."

"It's not a theory, it's quantum physics. Material particles have wave nature and can diffract around objects. If you send a beam of electrons at a barrier, they'll go around it to the left and to the right, at the same time." She's really getting into this, and she doesn't even notice the cat sunning itself in the yard across the street. "So, I'll just make use of my wave nature, and go around both sides of the pond."

"And where does running headfirst into a tree come in?"

"Oh, well." She looks a little sheepish. "I thought I would try it out on something smaller first. I got a good running start, and I was just about to go around when you stopped me."

"Ah. Like I said, an interesting theory. It won't work, you know."

"You're not going to try to claim I don't have wave nature, are you? Because I do. It's in your physics books."

"No, no, you've got wave nature, all right. You've also got Buddha nature—"

"I'm an enlightened dog!"

"—which will do you about as much good. You see, a tree is big, and your wavelength is small. At walking speed, a twenty-kilogram dog like you has a wavelength of about 10^{-35} meters. You need your wavelength to be comparable to the size of the tree—maybe ten centimeters—in order to diffract around it, and you're thirty-four orders of magnitude off."

"I'll just change my wavelength by changing my momentum. I can run very fast."

"Nice try, but the wavelength gets *shorter* as you go faster. To get your wavelength up to the millimeter or so you'd need to diffract around a tree, you'd have to be moving at 10^{-30} meters per second, and that's impossibly slow. It would take a billion years to cross the nucleus of an atom at that speed, which is way too slow to catch a bunny."

"So, you're saying I need a new plan?"

"You need a new plan."

Her tail droops, and we walk in silence for a few seconds. "Hey," she says, "can you help me with my new plan?"

"I can try."

"How do I use my Buddha nature to go around both sides of the pond at the same time?"

I really can't think of anything to say to that, but a flash of gray fur saves me. "Look! A squirrel!" I say.

"Oooooh!" And we're off in pursuit.

Quantum physics has many strange and fascinating aspects, but the discovery that launched the theory was **particle-wave duality**, or the fact that both light and matter have particle-like and wavelike properties at the same time. A beam of light, which is generally thought of as a wave, turns out to behave like a stream of particles in some experiments. At the same time, a beam of electrons, which is generally thought of as a stream of particles, turns out to behave like a wave in some experiments. Particle and wave properties seem to be contradictory, and yet everything in the universe somehow manages to be both a particle and a wave.

The discovery in the early 1900s that light behaves like a particle is the launching point for all of quantum mechanics. In this chapter, we'll describe the history of how physicists discovered this strange duality. In order to appreciate just what a strange development this is, though, we need to talk about the particles and waves that we see in everyday life.

PARTICLES AND WAVES AROUND YOU: CLASSICAL PHYSICS

Everybody is familiar with the behavior of material particles. Pretty much all the objects you see around you—bones, balls, squeaky toys—behave like particles in the classical sense, with their motion determined by classical physics. They have different shapes, but you can predict their essential motion by imagining each as a small, featureless ball with some mass—a particle—and applying Newton's laws of motion.* A tennis ball and a long bone tumbling end over end look very different in flight, but if they're thrown in the same direction with the same speed, they'll land in the same place, and you can predict that place using classical physics.

A particle-like object has a definite position (you know right where it is), a definite velocity (you know how fast it's moving, and in what direction), and a definite mass (you know how big it is). You can multiply the mass and velocity together, to find the **momentum**. A great big Labrador retriever has more momentum than a little French poodle when they're both moving at the same speed, and a fast-moving border collie has more momentum than a waddling basset hound of the same mass. Momentum determines what will happen when two particles collide. When a moving object hits a stationary one, the moving object will slow down, losing momentum, while the stationary object will speed up, gaining momentum.

The other notable feature of particles is something that seems almost too obvious to mention: particles can be counted. When you have some collection of objects, you can look at them and determine exactly how many of them you have—one bone, two squeaky toys, three squirrels under a tree in the backyard.

*Sir Isaac Newton, of the falling apple story, set forth three laws of motion that govern the behavior of moving objects. The first law is the principle of inertia, that objects at rest tend to remain at rest, and objects in motion tend to remain in motion unless acted on by an external force. The second law quantifies the first, and is usually written as the equation $F = ma$, force equals mass times acceleration. The third law says that for every action there is an equal and opposite reaction—a force of equal strength in the opposite direction. These three laws describe the motion of macroscopic objects at everyday speeds, and form the core of classical physics.

Waves, on the other hand, are slipperier. A wave is a moving disturbance in something, like the patterns of crests and troughs formed by water splashing in a backyard pond. Waves are spread out over some region of space by their nature, forming a pattern that changes and moves over time. No physical objects move anywhere—the water stays in the pond—but the pattern of the disturbance changes, and we see that as the motion of a wave.

If you want to understand a wave, there are two ways of looking at it that provide useful information. One is to imagine taking a snapshot of the whole wave, and looking at the pattern of the disturbance in space. For a single simple wave, you see a pattern of regular peaks and valleys, like this:

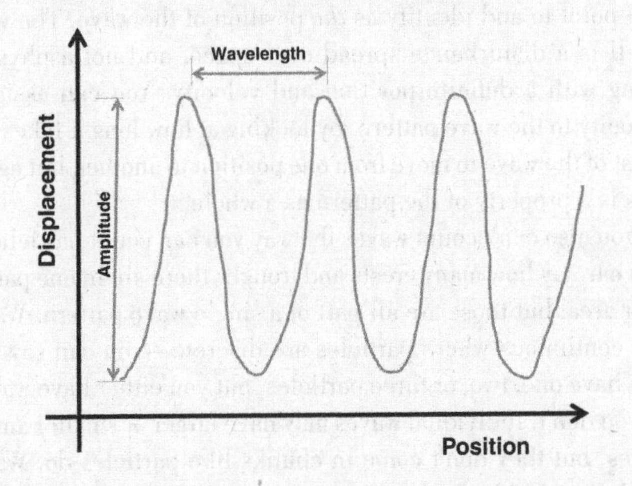

As you move along the pattern, you see the medium moving up and down by an amount called the "amplitude" of the wave. If you measure the distance between two neighboring crests of the wave (or two troughs), you've measured the "wavelength," which is one of the numbers used to describe a wave.

The other thing you can do is to look at one little piece of the wave pattern, and watch it for a long time—imagine watching a duck bobbing up and down on a lake, say. If you watch carefully, you'll see that the disturbance gets bigger and smaller in a very regular way—sometimes the duck is higher up, sometimes lower down—and makes a pattern in time very much like the pattern in space. You can measure how often the wave repeats itself in a given amount of time—how many times the duck reaches its maximum height in a minute, say—and that gives you the "frequency" of the wave, which is another critical number used to describe the wave. Wavelength and frequency are related to each other—longer wavelengths mean lower frequency, and vice versa.

You can already see how waves are different from particles: they don't have a position. The wavelength and the frequency describe the pattern as a whole, but there's no single place you can point to and identify as *the* position of the wave. The wave itself is a disturbance spread over space, and not a physical thing with a definite position and velocity. You can assign a velocity to the wave pattern, by looking at how long it takes one crest of the wave to move from one position to another, but again, this is a property of the pattern as a whole.

You also can't count waves the way you can count particles—you can say how many crests and troughs there are in one particular area, but those are all part of a single wave pattern. Waves are continuous where particles are discrete—you can say that you have one, two, or three particles, but you either have waves, or you don't. Individual waves may have larger or smaller amplitudes, but they don't come in chunks like particles do. Waves don't even add together in the same way that particles do—sometimes, when you put two waves together, you end up with a bigger wave, and sometimes you end up with no wave at all.

Imagine that you have two different sources of waves in the same area—two rocks thrown into still water at the same time, for example. What you get when you add the two waves together depends on how they line up. If you add the two waves together such that the crests of one wave fall on top of the crests of the other, and the troughs of one wave fall in the troughs of the other (such waves are called "in phase"), you'll get a larger wave than either of the two you started with. On the other hand, if you add two waves together such that the crests of one wave fall in the troughs of the other and vice versa ("out of phase"), the two will cancel out, and you'll end up with no wave at all.

This phenomenon is called *interference,* and it's perhaps the most dramatic difference between waves and particles.

"I don't know . . . that's pretty weird. Do you have any other examples of interference? Something more . . . doggy?"

"No, I really don't. That's the point—waves are dramatically different than particles. Nothing that dogs deal with on a regular basis is all that wavelike."

"How about, 'Interference is like when you put a squirrel in the backyard, and then you put a dog in the backyard, and a minute later, there's no more squirrel in the backyard.'"

"That's not interference, that's prey pursuit. Interference is more like putting a squirrel in the backyard, then putting a second squirrel in the backyard one second later, and finding that you have no squirrels at all. But if you wait *two* seconds before putting in the second squirrel, you find four squirrels."

"Okay, that's just weird."

"That's my point."

"Oh. Well, good job, then. Anyway, why are we talking about this?"

"Well, you need to know a few things about waves in order to understand quantum physics."

"Yeah, but this just sounds like math. I don't like math. When are we going to talk about physics?"

"We are talking about physics. The whole point of physics is to use math to describe the universe."

"I don't want to describe the universe, I want to catch squirrels."

"Well, if you know how to describe the universe with math, that can help you catch squirrels. If you have a mathematical model of where the squirrels are now, and you know the rules governing squirrel behavior, you can use your model to predict where they'll be later. And if you can predict where they'll be later . . ."

"I can catch squirrels!"

"Exactly."

"All right, math is okay. I still don't see what this wave stuff is for, though."

"We need it to explain the properties of light and sound waves, which is the next bit."

WAVES IN EVERYDAY LIFE: LIGHT AND SOUND

We deal with two kinds of waves in everyday life: light and sound. Though these are both examples of wave phenomena, they appear to behave very differently. The reasons for those differences will help shed some light (pardon the pun) on why it is that we don't see dogs passing around both sides of a tree at the same time.

Sound waves are pressure waves in the air. When a dog barks, she forces air out through her mouth and sets up a vibration that travels through the air in all directions. When it reaches another dog, that sound wave causes vibrations in the second dog's eardrums, which are turned into signals in the brain that are processed as sound, causing the second dog to bark, producing more waves, until nearby humans get annoyed.

Light is a different kind of wave, an oscillating electric and magnetic field that travels through space—even the emptiness of outer space, which is why we can see distant stars and galaxies. When light waves strike the back of your eye, they get turned into signals in the brain that are processed to form an image of the world around you.

The most striking difference between light and sound in everyday life has to do with what happens when they encounter an obstacle. Light waves travel only in straight lines, while sound waves seem to bend around obstacles. This is why a dog in the dining room can hear a potato chip hitting the kitchen floor, even though she can't see it.

The apparent bending of sound waves around corners is an example of **diffraction**, which is a characteristic behavior of waves encountering an obstacle. When a wave reaches a barrier with an opening in it, like the wall containing an open door from the kitchen into the dining room, the waves passing through the opening don't just keep going straight, but fan out over a range of different directions. How quickly they spread depends on the wavelength of the wave and the size of the opening through

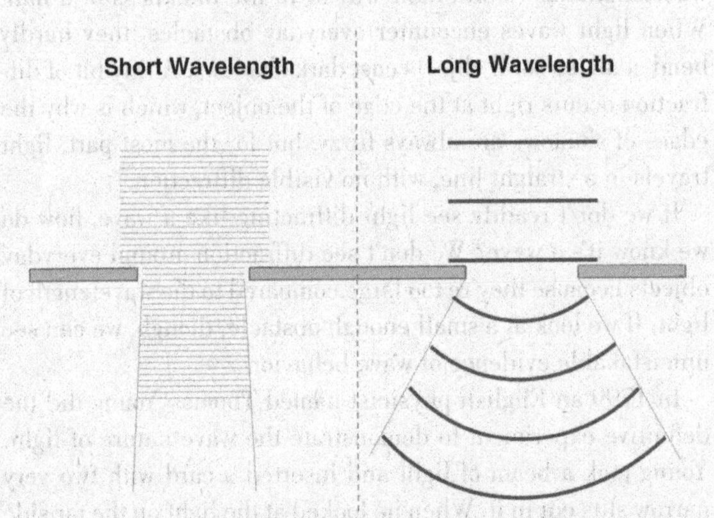

On the left, a wave with a short wavelength encounters an opening much larger than the wavelength, and the waves continue more or less straight through. On the right, a wave with a long wavelength encounters an opening comparable to the wavelength, and the waves diffract through a large range of directions.

which they travel. If the opening is much larger than the wavelength, there will be very little bending, but if the opening is comparable to the wavelength, the waves will fan out over the full available range.

Similarly, if sound waves encounter an obstacle like a chair or a tree, they will diffract around it, provided the object is not too much larger than the wavelength. This is why it takes a large wall to muffle the sound of a barking dog—sound waves bend around smaller obstacles, and reach people or dogs behind them.

Sound waves in air have a wavelength of a meter or so, close to the size of typical obstacles—doors, windows, pieces of furniture. As a result, the waves diffract by a large amount, which is why we can hear sounds even around tight corners.

Light waves, on the other hand, have a very short wavelength—less than a thousandth of a millimeter. A hundred wavelengths of visible light will fit in the thickness of a hair. When light waves encounter everyday obstacles, they hardly bend at all, so solid objects cast dark shadows. A tiny bit of diffraction occurs right at the edge of the object, which is why the edges of shadows are always fuzzy, but for the most part, light travels in a straight line, with no visible diffraction.

If we don't readily see light diffracting like a wave, how do we know it's a wave? We don't see diffraction around everyday objects because they're too large compared to the wavelength of light. If we look at a small enough obstacle, though, we can see unmistakable evidence of wave behavior.

In 1799 an English physicist named Thomas Young did the definitive experiment to demonstrate the wave nature of light. Young took a beam of light and inserted a card with two very narrow slits cut in it. When he looked at the light on the far side of the card, he didn't just see an image of the two slits, but rather a large pattern of alternating bright and dark spots.

Young's double-slit experiment is a clear demonstration of the diffraction and interference of light waves. The light pass-

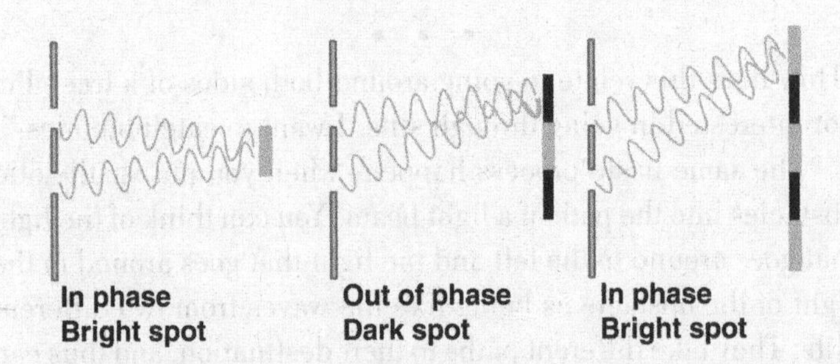

An illustration of double-slit diffraction. On the left, the waves from two different slits travel exactly the same distance, and arrive in phase to form a bright spot. In the center, the wave from the lower slit travels an extra half-wavelength (darker line), and arrives out of phase with the wave from the upper slit. The two cancel out, forming a dark spot in the pattern. On the right, the wave from the lower slit travels a full extra wavelength, and again adds to the wave from the upper slit to form a bright spot.

ing through each of the slits diffracts out into a range of different directions, and the waves from the two slits overlap. At any given point, the waves from the two slits have traveled different distances, and have gone through different numbers of oscillations. At the bright spots, the two waves are in phase, and add together to give light that is brighter than light from either slit by itself. At the dark spots, the waves are out of phase, and cancel each other out.

Prior to Young's experiment, there had been a lively debate about the nature of light, with some physicists claiming that light was a wave, and others (including Newton) arguing that light was a stream of tiny particles. Interference and diffraction are phenomena that only happen with waves, though, so after Young's experiment (and subsequent experiments by the French physicist Augustin Fresnel), everybody was convinced that light was a wave. Things stayed that way for about a hundred years.

• • •

"How does this relate to going around both sides of a tree? I'm not interested in going through slits, I want to catch bunnies."

"The same basic process happens when you put small solid obstacles into the path of a light beam. You can think of the light that goes around to the left and the light that goes around to the right of the obstacle as being like the waves from two different slits. They take different paths to their destination, and thus can be either in phase or out of phase when they arrive. You get a pattern of bright and dark spots, just like when you use slits."

"Oh. I guess that makes sense. So, I just need to get the bunnies to stand at the spots where I'm in phase with me?"

"No, because of the wavelength thing. We'll get to that in a minute. I need to talk about particles, first."

"Okay. I can be patient. As long as it doesn't take too long."

THE BIRTH OF THE QUANTUM: LIGHT AS A PARTICLE

The first hint of a problem with the wave model of light came from a German physicist named Max Planck in 1900. Planck was studying the **thermal radiation** emitted by all objects. The emission of light by hot objects is a very common phenomenon (the best-known example is the red glow of a hot piece of metal), and something so common seems like it ought to be easy to explain. By 1900, though, the problem of explaining how much light of different colors was emitted (the "spectrum" of the light) had thus far defeated the best physicists of the nineteenth century.

Planck knew that the spectrum had a very particular shape, with lots of light emitted at low frequencies and very little at high frequencies, and that the peak of the spectrum—the frequency at which the light emitted is brightest—depends only on the object's temperature. He had even discovered a formula to describe the characteristic shape of the spectrum, but was stymied when he tried to find a theoretical justification for the formula. Every method he tried predicted much more light at high frequencies than was observed. In desperation, he resorted to a mathematical trick to get the right answer.

Planck's trick was to imagine that all objects contained fictitious "oscillators" that emit light only at certain frequencies. Then he said that the amount of energy (E) associated with each oscillator was related to the frequency of the oscillation (f) by a simple formula:

$$E = hf$$

where h is a constant. When he first made this odd assumption, Planck thought he would use it just to set up the problem, and then use a common mathematical technique to get rid of the imaginary oscillators and this extra constant h. Much to his surprise, though, he found that his results made sense only if he kept the oscillators around—if h had a very small but nonzero value.

Today, h is known as **Planck's constant** in his honor, and has the value 6.626×10^{-34} kg m^2/s (that's 0.000000000000000000000000000000000006626 kg m^2/s). It's a very small number indeed, but definitely not zero.

Planck's trick amounts to treating light, which physicists thought of as a continuous wave, as coming in discrete chunks, like particles. Planck's "oscillators" could only emit light in discrete units of brightness. This is a little like imagining a pond where waves can only be one, two, or three centimeters high, never one and a half or two and a quarter. Everyday waves don't work that way, but that's what Planck's mathematical model requires.

These "oscillators" are also what puts the "quantum" in "quantum physics." Planck referred to the specific levels of energy in his oscillators as "quanta" (the plural of "quantum," from the

Latin word for "how much"), so an oscillator at a given frequency might contain one quantum (one unit of energy, hf), two quanta, three quanta, and so on, but never one and a half or two and a quarter. The name for the steps stuck, and came to be applied to the entire theory that grew out of Planck's desperate trick.

Though he's often given credit for inventing the idea of light quanta, Planck never really believed that light came in discrete quanta, and he always hoped that somebody would find a clever way to derive his formula without resorting to trickery.

The first person to talk seriously about light as a quantum particle was Albert Einstein in 1905, who used it to explain the **photoelectric effect**. The photoelectric effect is another physical effect that seems like it ought to be simple to describe: when you shine light on a piece of metal, electrons come out. This forms the basis for simple light sensors and motion detectors: light falling on a sensor knocks electrons out of the metal, which then flow through a circuit. When the amount of light hitting the sensor changes, the circuit performs some action, such as turning lights on when it gets dark, or opening doors when a dog passes in front of the sensor.

The photoelectric effect ought to be readily explained by thinking of light as a wave that shakes atoms back and forth until electrons come out, like a dog shaking a bag of treats until they fly all over the kitchen. Unfortunately, the wave model comes out all wrong: it predicts that the energy of the electrons leaving the atoms should depend on the intensity of the light—the brighter the light, the harder the shaking, and the faster the bits flying away should move. In experiments, though, the energy of the electrons doesn't depend on the intensity at all. Instead, the energy depends on the frequency, which the wave model says shouldn't matter. At low frequencies, you never get any electrons no matter how hard you shake, while at high frequency, even gentle shaking produces electrons with a good deal of energy.

• • •

"Physicists are silly."

"I beg your pardon?"

"Well, any dog knows *that*. When you get a bag with treats in it, you always shake it as fast as you can, as hard as you can. That's how you get the treats out."

"Yes, well, what can I say? Dogs have an excellent intuitive grasp of quantum theory."

"Thank you. We're cute, too."

"Of course, the point of physics is to understand *why* the treats come out when they do."

"Maybe for you. For dogs, the point is to get the treats."

Einstein explained the photoelectric effect by applying Planck's formula to light itself. Einstein described a beam of light as a stream of little particles, each with an energy equal to Planck's constant multiplied by the frequency of the light wave (the same rule used for Planck's "oscillators"). Each **photon** (the name now given to these particles of light) has a fixed amount of energy it can provide, depending on the frequency; and some minimum amount of energy is required to knock an electron loose. If the energy of a single photon is more than the minimum needed, the electron will be knocked loose, and carry the rest of the photon's energy with it. The higher the frequency, the higher the single photon energy and the more energy the electrons have when they leave, exactly as the experiments show. If the energy of a single photon is lower than the minimum energy for knocking an electron out, nothing happens, explaining the lack of electrons at low frequencies.*

Describing light as a particle was a hugely controversial idea in 1905, as it overturned a hundred years' worth of physics and

*You might wonder why you can't put together two low-energy photons to provide enough energy to free an electron. This would require two photons to hit the same electron at the same instant, and that almost never happens.

requires a very different view of light. Rather than a continuous wave, like water poured into a dog's bowl, light has to be thought of as a stream of discrete particles, like a scoop of kibble poured into a bowl. And yet each of those particles still has a frequency associated with it, and somehow they add up to give an interference pattern, just like a wave.

Other physicists in 1905 found this deeply troubling, and Einstein's model took a while to gain acceptance. The American physicist Robert Millikan hated Einstein's idea, and performed a series of extremely precise photoelectric effect experiments in 1916 hoping to prove Einstein wrong.* In fact, his results confirmed Einstein's predictions, but even that wasn't enough to get the photon idea accepted. Wide acceptance of the photon picture didn't come until 1923, when Arthur Holly Compton did a famous series of experiments with X-rays that demonstrated unmistakably particle-like behavior from light: he showed that photons carry momentum, and this momentum is transferred to other particles in collisions.

If you take the Planck formula for the energy of a single photon, and combine it with equations from Einstein's special relativity, you find that a single photon of light ought to carry a small amount of momentum, given by the formula:

$$p = h/\lambda$$

where p is the symbol for momentum and λ is the wavelength of the light.

• • •

*Millikan thought the Einstein model lacked "any sort of satisfactory theoretical foundation," and described its success as "purely empirical," which is pretty nasty by physics standards. Ironically, those quotes are from the first paragraph of the paper in which he conclusively confirms the predictions of the theory.

"I thought you said there wasn't any relativity in this book?"

"I said the book isn't *about* relativity. That's not the same thing. Some ideas from relativity are important to quantum physics, as well."

"What's relativity got to do with this, though?"

"Well, what relativity says is that because a photon has some energy, it must have some momentum, even though it doesn't have any mass."

"So . . . it's an $E = mc^2$ thing?"

"Not exactly, but it's similar. Photons have momentum because of their energy in the same way that objects have energy because of their mass. And nice job dropping an equation in there."

"Please. Even inferior dogs know $E = mc^2$. And I am an *exceptional* dog."

A photon with a small wavelength has a lot of momentum, while a photon with a large wavelength has very little. That means that the interaction between a photon of light and a stationary object ought to look just like a collision between two particles: the stationary object gains some energy and momentum, and the moving photon loses some energy and momentum. We don't notice this because the momentum involved is tiny—Planck's constant is a very small number—but if we look at an object with a very small mass, like an electron, and photons with a very short wavelength (and thus a relatively high momentum), we can detect the change in momentum.

In 1923, Compton bounced X-rays with an initial wavelength of 0.0709 nanometers* off a solid target (X-rays are just light with an exceptionally short wavelength, compared to about 500 nm for visible light). When he looked at the X-rays that scattered off the target, he found that they had longer wavelengths, indicating that they had lost momentum (X-rays bouncing off

*A nanometer is 10^{-9} m, or one billionth of a meter (0.000000001 m).

at 90 degrees from their original direction had a wavelength of 0.0733 nm, for example). This loss of momentum is exactly what should happen if light is a particle: when an X-ray photon comes in and hits a more or less stationary electron in a target, it gives up some of its momentum to the electron, which starts moving. After the collision, the photon has less momentum, and thus a longer wavelength, exactly as Compton observed.

The amount of momentum lost also depends on the angle at which the photon bounces off—a photon that glances off an electron doesn't lose very much momentum, while one that bounces almost straight back loses a lot. Compton measured the wavelength at many different angles, and his results exactly fit the theoretical prediction, confirming that the shift was from collisions with electrons, and not some other effect.

Einstein, Millikan, and Compton all won Nobel prizes for demonstrating the particle nature of light. Taken together, Millikan's photoelectric effect experiments and Compton's scattering experiments were enough to get most physicists to accept the idea of light as being made up of a stream of particles.*

As strange as the idea of light as a particle was, though, what came next was even stranger.

*A few die-hard theorists still resisted the idea of photons, because even the Compton effect can be explained without photons, though it's very complicated. The last resistance collapsed in 1977, when incontrovertible proof of the existence of photons was provided in an experiment by Kimble, Dagenais, and Mandel that looked at the light emitted by single atoms. The seventy-two-year gap between Einstein's proposal and its final acceptance tells you something about the stubbornness of physicists confronted with a new idea. It can be as difficult to separate a physicist from a cherished model as it is to drag a dog away from a well-chewed bone.

INTERFERING ELECTRONS: PARTICLES AS WAVES

Also in 1923, a French Ph.D. student named Louis Victor Pierre Raymond de Broglie* made a radical suggestion: he argued that there ought to be symmetry between light and matter, and so a material particle such as an electron ought to have a wavelength. After all, if light waves behave like particles, shouldn't particles behave like waves?

De Broglie suggested that just as a photon has a momentum determined by its wavelength, a material object like an electron should have a wavelength determined by its momentum:

$$\lambda = h/p$$

which is just the formula for the momentum of a photon (page 24) turned around to give the wavelength. The idea has a certain mathematical elegance, which was appealing to theoretical physicists even in 1923, but it also seems like patent nonsense—solid objects show no sign of behaving like waves. When de Broglie presented his idea as part of his Ph.D. thesis defense, nobody knew what to make of it. His professors weren't even sure whether to give him the degree or not, and resorted to showing his thesis to Einstein. Einstein proclaimed it brilliant, and de Broglie got his degree, but his idea of electrons as waves had little support until two experiments in the late 1920s showed incontrovertible proof that electrons behaved like waves.

*The proper pronunciation of Louis de Broglie's surname (his collection of names reflects his aristocratic background—he was the 7th Duc de Broglie) is the source of much confusion among American physicists. I've heard "de-BRO-lee," "de-BRO-glee," and "de-BROY-lee," among others. The correct French pronunciation is apparently something close to "de-BROY," only with a gargly sort of sound to the vowel that you need to be French to make.

In 1927, two American physicists, Clinton Davisson and Lester Germer, were bouncing electrons off a surface of nickel, and recording how many bounced off at different angles. They were surprised when their detector picked up a very large number of electrons bouncing off at one particular angle. This mysterious result was eventually explained as the wavelike diffraction of the electrons bouncing off different rows of atoms in their nickel target. The beam of electrons penetrated some distance into the nickel, and part of the beam bounced off the first row of atoms in the nickel crystal,* while other parts bounced off the second, and the third, and so on. Electrons reflecting from all these different rows of atoms behaved like waves. The waves that bounced off atoms deeper in the crystal traveled farther on the way out than the ones that bounced off atoms closer to the surface. These waves interfered with one another, like light waves passing through the different slits in Young's experiment (though with many slits, not just two). Most of the time, the reflected waves were out of phase and canceled one another out. For certain angles, though, the extra distance traveled was exactly right for the waves to add in phase and produce a bright spot, which Davisson and Germer detected as a large increase in the number of electrons reflected at that angle. The de Broglie formula for assigning a wavelength to the electron predicts the Davisson and Germer result perfectly.†

*"Crystal," to a physicist, refers to any solid with a regular and orderly arrangement of atoms in it. This includes the clear and sparkly things that we normally associate with the word, but also a lot of metals and other substances.

†Ironically, Davisson and Germer succeeded only because they broke a piece of their apparatus. They didn't see any diffraction in the first experiments they did, because their nickel target was made up of many small crystals, each producing a different interference pattern, and the bright spots from the different patterns ran together. Then they accidentally let air into their vacuum system. In the process of repairing the damage, they melted the target, which recrystallized into a single large crystal, producing a single, clear

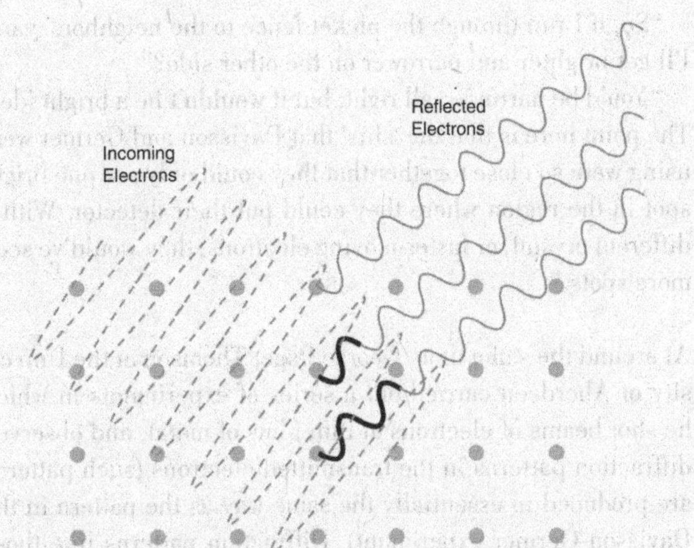

Diffraction of electrons off a crystal of nickel. An incoming electron beam (dashed line) passes into a regular crystal of atoms, and bits of the wave (individual electrons) reflect off different atoms in the crystal. Electrons reflected from deeper in the crystal travel a longer distance on the way out (darker line), but for certain angles, that distance is a multiple of a full wavelength, and the waves leaving the crystal add in phase to give the bright spot seen by Davisson and Germer.

• • •

"Wait, how does that work? If there are lots of slits, shouldn't there be lots of spots?"

"Not really. When you add the waves together, you still get a pattern of bright and dark spots, but as you use more slits, the bright spots get brighter and narrower, and the dark spots get darker and wider."

---

diffraction pattern. Sometimes, the luckiest thing a physicist can do is to break something important.

"So, if I run through the picket fence to the neighbors' yard, I'll get brighter and narrower on the other side?"

"You'd be narrower, all right, but it wouldn't be a bright idea. The point here is that the 'slits' that Davisson and Germer were using were so close together that they could only see one bright spot in the region where they could put their detector. With a different crystal, or faster-moving electrons, they would've seen more spots."

At around the same time, George Paget Thomson at the University of Aberdeen carried out a series of experiments in which he shot beams of electrons at thin films of metal, and observed diffraction patterns in the transmitted electrons (such patterns are produced in essentially the same way as the pattern in the Davisson-Germer experiment). Diffraction patterns like those seen by Davisson and Germer and Thomson are an unmistakable signature of wave behavior, as Thomas Young showed in 1799, so their experiments provided proof that de Broglie was right, and electrons have wave nature. De Broglie won the Nobel Prize in Physics in 1929 for his prediction, and Davisson and Thomson shared a Nobel Prize in 1937 for demonstrating the wave nature of the electron.*

Following the experiments of Davisson and Germer and Thomson, scientists showed that all subatomic particles behave like waves: beams of protons and neutrons will diffract off samples of atoms in exactly the same way that electrons do. In fact, neutron diffraction is now a standard tool for determining the structure of materials at the atomic level: scientists can deduce how atoms are arranged by looking at the interference patterns

*In one of the great bits of Nobel trivia, Thomson's father, J. J. Thomson of Cambridge, won the 1906 Nobel Prize in Physics for demonstrating the particle nature of the electron. This presumably led to some interesting dinner-table conversation in the Thomson household.

that result when a beam of neutrons bounces off their sample. Knowing the structure of materials at the atomic level allows materials scientists to design stronger and lighter materials for use in cars, planes, and space probes. Neutron diffraction can also be used to determine the structure of biological materials like proteins and enzymes, providing critical information for scientists searching for new drugs and medical treatments.

EVERYTHING IS MADE OF WAVES: INTERFERENCE OF MOLECULES

So, if all material objects are made up of particles with wave properties, why don't we see dogs diffracting around trees? If a beam of electrons can diffract off two rows of atoms, why can't a dog run around both sides of a tree to trap a bunny on the far side? The answer is the wavelength: as with the sound and light waves discussed earlier, the dramatically different behavior of dogs and electrons encountering obstacles is explained by the difference in their wavelengths. The wavelength is determined by the momentum, and a dog has a lot more momentum than an electron.

The wavelength of a material object is given by Planck's constant divided by the momentum, which is mass multiplied by velocity. Planck's constant is a tiny number, but so is the mass of an electron—about 10^{-30} kilograms, or 0.0000000000000000 00000000000001 kg. Davisson and Germer's electrons, moving at the brisk speed of six million meters per second, had a wavelength of about a tenth of a nanometer (0.0000000001 m). That's extremely small, but it's about half the distance between two nickel atoms, just right for seeing diffraction (just like sound waves with half-meter wavelengths will readily diffract through one-meter-wide doors).

The wavelength of a 50-pound (about 20 kg) dog out for a stroll, on the other hand, is about 10^{-35} meters (0.00000000000

0000000000000000000000001 m), or a millionth of a billionth of a billionth of the wavelength of Davisson and Germer's electrons. How does that compare to the size of a tree? Well, a dog's wavelength compared to the distance between two atoms is like the distance between two atoms compared to the diameter of the solar system. There's no chance of seeing the wave associated with a dog diffract off a crystal of nickel, let alone pass around both sides of a tree at the same time.

There's a lot of room between a beam of electrons and a dog, though, so what is the biggest material object that has been shown to have observable wave nature?

In 1999, a research group at the University of Vienna headed

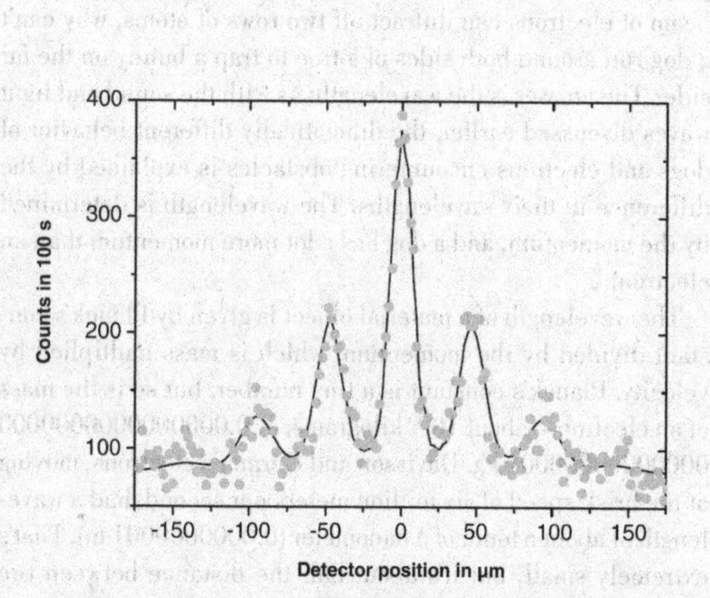

The interference pattern produced by a beam of molecules passing through an array of narrow slits. The extra lumps to either side of the central peak are the result of diffraction and interference of the molecules passing through the slits. (Reprinted with permission from O. Nairz, M. Arndt, and A. Zeilinger, Am. J. Phys. *71, 319 [2003]. Copyright 2003, American Association of Physics Teachers.)*

by Dr. Anton Zeilinger observed diffraction and interference with molecules consisting of 60 carbon atoms bound together into a shape like a tiny soccer ball, each with a mass about a million times that of an electron. They shot these soccer-ball-shaped molecules toward a detector, and when they looked at the distribution of molecules downstream, they saw a single narrow beam. Then they sent the beam through a silicon wafer with a collection of very small slits cut into it, and looked at the distribution of molecules on the far side of the slits. With the slits in place, the initial narrow peak grew broader, with distinct "lumps" to either side. Those lumps, like the bright and dark spots seen by Thomas Young shining light through a double slit, or the electron diffraction peaks seen by Davisson and Germer, are an unmistakable signature of wave behavior. Molecules passing through the slits spread out and interfere with one another, just like light waves.

In subsequent experiments, the Zeilinger group demonstrated the diffraction of even larger molecules, adding 48 fluorine atoms to each of their original 60-carbon-atom molecules. These molecules have a mass about three million times the mass of one electron, and stand as the current record for the most massive object whose wave nature has been observed directly.

As the mass of a particle increases, its wavelength gets shorter and shorter, and it gets harder and harder to see wave effects directly. This is why nobody has ever seen a dog diffract around a tree; nor are we likely to see it any time soon. In terms of physics, though, a dog is nothing but a collection of biological molecules, which the Zeilinger group has shown have wave properties. So, we can say with confidence that a dog has wave nature, just the same as everything else.

"So, which are they *really*?"

"What do you mean?"

"Well, are electrons really particles acting like waves, or are photons waves acting like particles?"

"You're asking the wrong questions. Or, rather, you're giving the wrong answers. The real answer is 'Door Number Three.' Electrons and photons are both examples of a third sort of object, which is neither just a wave nor just a particle, but has some wave properties and some particle properties at the same time."

"So, it's like a squirunny?"

"A what?"

"A critter that's something like a squirrel, and something like a bunny. A squirunny."

"I prefer 'quantum particle,' but I guess that's the basic idea. Everything in the universe is built of these quantum particles."

"That's pretty weird."

"Oh, that's just the beginning of the weird stuff . . ."

CHAPTER 2

Where's My Bone? The Heisenberg Uncertainty Principle

I'm grading papers on the couch when Emmy comes into the room, looking concerned. "What's the matter?" I ask.

"I can't find my bone," she says. "Do you know where my bone is?"

"I have no idea where your bone is," I say, "but I can tell you exactly how fast it's moving."

There's silence in response, and when I look up, she's staring at me blankly.

"It's a physics joke," I explain, because that always makes things funnier. "You know, Heisenberg's uncertainty principle? The uncertainty in the position of an object multiplied by the uncertainty in the momentum is greater than Planck's constant over four pi? Which means that when one uncertainty is small, the other must be very large."

Now she's glaring at me, almost growling. "Stop doing that!" she says.

"What? It's not all that funny, but it wasn't that bad."

"It's your fault that I can't find my bone."

"How is it my fault?"

"You went and measured how fast it's moving, and the position got all uncertain. And now I can't find my bone."

"That's not what happened," I say. "The uncertainty principle doesn't work like that."

"Yes it does. You just said. You know how fast my bone is moving, and now I can't find it."

"First of all, that was a joke. I didn't really measure the velocity of your bone. Second, that's a slightly mistaken view of the uncertainty principle. It's not just that measurement changes the state of the system, it's that what we *can* measure is limited by the fact that position and momentum are undefined until we measure them."

She looks puzzled. "I don't see the difference."

"Well, in the picture where you attribute everything to the effects of measurement, you implicitly assume that whatever you're measuring has some definite and well-defined properties, and the uncertainty in those values arises only from perturbations that occur through the act of measuring them. That's not what happens, though—in quantum theory, there are no definite values for those quantities. They're not uncertain because of limits on your measurement, they're uncertain because they are not defined, and they can't *be* defined, due to the quantum nature of reality."

"Oh." She looks thoughtful for a moment, then resumes glaring. "I think you lost my bone, and you're just trying to weasel out of this by being all confusing."

"No, that's really how the theory works. It's a moot point, though, since even if I had perturbed the position of your bone by measuring its velocity, there's no way that would've prevented you from finding it."

"Yeah? Why not?"

"Well, because the uncertainty involved would be tiny. I mean, your bone has a mass of a couple hundred grams, and if I measured its velocity to within one millimeter per second, that would give an uncertainty in position of only about 10^{-31}

meters. That's a trillionth of the size of a proton—you'd never even notice that."

"Yeah? Well, where's my bone, smart guy?"

"I don't know. Did you look under the TV cabinet? Sometimes it gets kicked under there."

She trots over to the TV, and sticks her nose under the cabinet. "Oooh! Here's my bone!" She paws at it for a minute, and eventually succeeds in knocking it out from under the cabinet. "I have a bone!" she announces proudly, and begins chewing it noisily, the uncertainty principle forgotten.

The Heisenberg **uncertainty principle** is probably the second most famous result from modern physics, after Einstein's $E = mc^2$ (the most famous result from relativity). Most people wouldn't know a wavefunction if they tripped over one, but almost everyone has heard of the uncertainty principle: it is impossible to know both the position and the momentum of an object perfectly at the same time. If you make a better measurement of the position, you necessarily lose information about its momentum, and vice versa.

In this chapter, we'll describe how the uncertainty principle arises from the particle-wave duality we've already discussed. The uncertainty principle is often presented as a statement that a measurement of a system changes the state of that system, and in this form, references to quantum uncertainty turn up in all sorts of places, from politics to pop culture to sports.* Ulti-

*To give you an idea of the breadth of subjects in which this shows up, in June 2008, Google turned up citations of the Heisenberg uncertainty principle in (among others) an article from the *Vermont Free Press* about traffic cameras, a *Toronto Star* article citing the influence of YouTube on underground artists, and a blog article about the Phoenix Suns of the NBA. Incidentally, all of these articles also use the uncertainty principle incorrectly—by the end of this chapter, you should hopefully understand it better than any of them.

mately, though, uncertainty has very little to do with the details of the measurement process. Quantum uncertainty is a fundamental limit on what *can* be known, arising from the fact that quantum objects have both particle and wave properties.

Uncertainty is also the first place where quantum physics collides with philosophy. The idea of fundamental limits to measurement runs directly counter to the goals and foundations of classical physics. Dealing with quantum uncertainty requires a complete rethinking of the basis of physics, and leads directly to the issues of measurement and interpretation in chapters 3 and 4.

HEISENBERG'S MICROSCOPE: SEMICLASSICAL ARGUMENTS

The traditional description of uncertainty as the act of measurement changing the state of the system is essentially based in classical physics, and was developed in the 1920s and '30s in order to convince classically trained physicists that quantum uncertainty needed to be taken seriously. This is what physicists call a **semiclassical argument**—the physics used is classical, with a few modern ideas added on. It's not the full picture, but it has the advantage of being readily comprehensible.

The idea behind the semiclassical treatment of uncertainty is familiar to any dog. Imagine you have a bunny in the yard whose position and velocity you would like to know very well. When you attempt to make a better determination of its position (by getting closer to it), you inevitably change its velocity by making it run away. No matter how slowly you creep up on it, sooner or later, it always takes off, and you never really have a good idea of both the position and the velocity.

An electron isn't a sentient being like a bunny, so it can't run off of its own accord, but a similar process takes place.

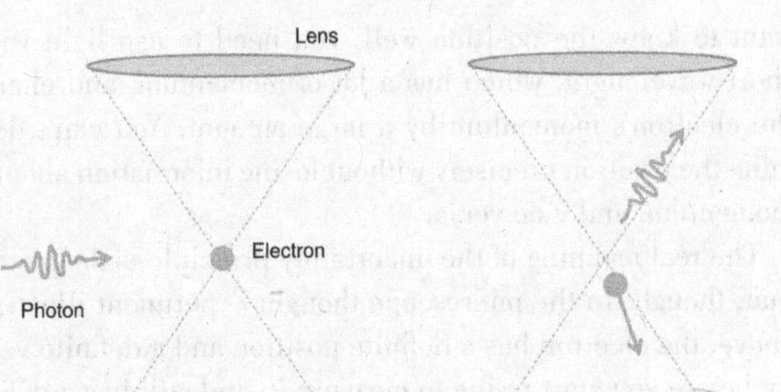

An incoming photon bounces off a stationary electron, and is collected by a microscope lens in order to measure the electron's position. In the collision, though, the electron acquires some momentum, leading to uncertainty in its momentum.

To measure the position of an electron, you need to do something to make it visible, such as bouncing a photon of light off it and viewing the scattered light through a microscope. But the photon carries momentum (as we saw in chapter 1 [page 24]), and when it bounces off the electron, it changes the momentum of the electron. The electron's momentum after the collision is uncertain, because the microscope lens collects photons over some range of angles, so you can't tell exactly which way it went.

You can make the momentum change smaller by increasing the wavelength of the light (decreasing the momentum that the photon has available to give to the electron), but when you increase the wavelength, you decrease the resolution of your microscope, and lose information about the position.* If you

*This is why scientists use electron microscopes to look at very small things: electron microscopes use electrons instead of visible light, and electrons have much shorter wavelengths than visible light.

want to know the position well, you need to use light with a short wavelength, which has a lot of momentum, and changes the electron's momentum by a large amount. You can't determine the position precisely without losing information about the momentum, and vice versa.

The real meaning of the uncertainty principle is deeper than that, though. In the microscope thought experiment illustrated above, the electron has a definite position and a definite velocity before you start trying to measure it, and still has a definite position and velocity after the measurement. You don't know what the position and velocity are, but they have definite values. In quantum theory, however, these quantities are not defined. Uncertainty is not a statement about the limits of measurement, it's a statement about the limits of reality. Asking for the precise position and momentum of a particle doesn't even make sense, because those quantities do not exist.

This fundamental uncertainty is a consequence of the dual nature of quantum particles. As we saw in the previous chapter, experiments have shown that light and matter have both particle-like and wavelike properties. If we're going to describe quantum particles mathematically—and physics is all about mathematical description of reality—we need to find some way of talking about these objects that allows them to have both particle and wave properties at the same time. We'll find that the only way is to have both the position and the momentum of the quantum particles be uncertain.

BUILDING A QUANTUM PARTICLE: PROBABILITY WAVES

The usual way of describing particles mathematically, dating from the late 1920s, is through quantum **wavefunctions**. The wavefunction for a particular object is a mathematical function that has some value at every point in the universe, and that value squared gives the probability of finding a particle at a given position at a given time. So the question we need to ask is, What sort of wavefunction gives a probability distribution that has both particle and wave properties?

Constructing a probability distribution for a classical particle is easy, and the result looks something like this:

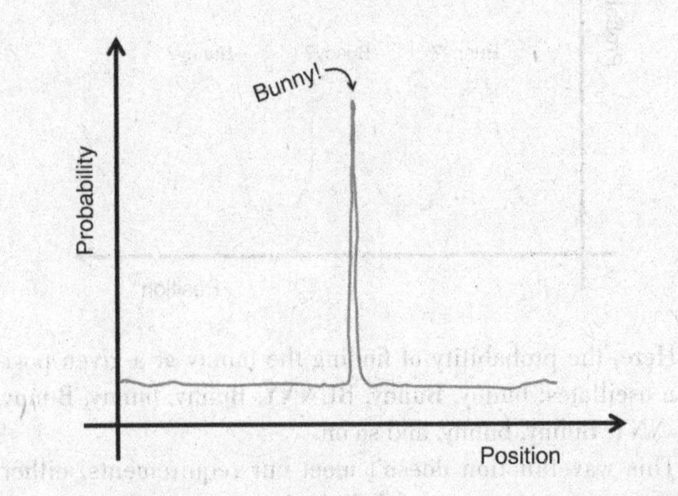

The probability of finding the object—say, that pesky bunny in the backyard—is zero everywhere except right at the well-defined position of the object. As you look across the yard, you see nothing, nothing, nothing, BUNNY!, nothing, nothing, nothing.

This wavefunction doesn't meet our requirements, though: it has a well-defined position, but it's just a single spike, and a spike does not have a wavelength. Remember, the wavelength corresponds to the momentum of the bunny, which is one of the quantities we're trying to describe, so it needs to have some value.

Well, then, how do we draw a probability distribution with an obvious wavelength? That's also easy to do, and it looks like this:

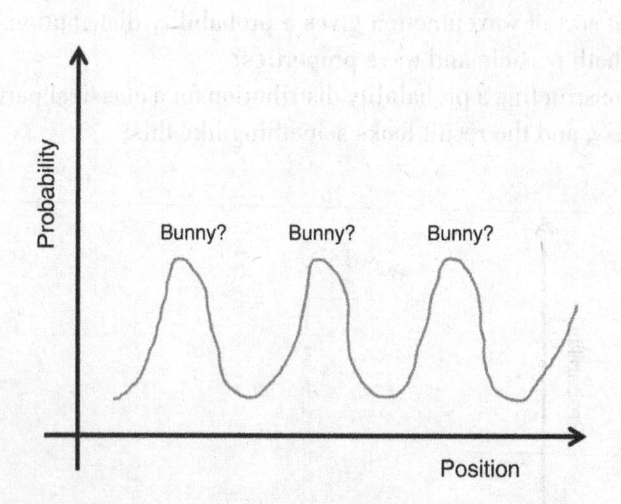

Here, the probability of finding the bunny at a given position oscillates: bunny, Bunny, BUNNY, Bunny, bunny, Bunny, BUNNY, Bunny, bunny, and so on.

This wavefunction doesn't meet our requirements, either. The wavelength is easy to define—just measure the distance between two points where the probability is largest—so we have a well-defined momentum, but we can't identify a specific position for the bunny. The bunny is spread out over the entire yard, with a good probability of finding it at lots of different places. There are places where the probability of seeing a bunny is low, but they don't account for much space.

What we need is a "wave packet," a wavefunction that combines particle and wave properties in a single probability distribution, like this:

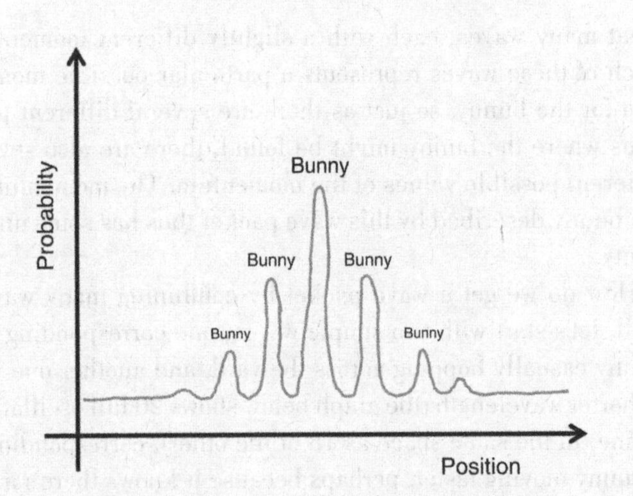

This wavefunction is what we're after: nothing, nothing, bunny, Bunny, BUNNY, Bunny, bunny, nothing, nothing. The bunny is very likely to be found in a small region of space, and the probability of finding it outside that region drops off to zero. Inside that region, we see oscillations in the probability, which allow us to measure a wavelength, and thus the momentum.

This wave packet has the particle and wave properties that we're looking for. As a consequence, it *also* has some uncertainty in both the position and momentum of the particle.

The uncertainty in position is immediately obvious on looking at the wave packet. The bunny can't be pinned down to a specific location, but there are several different positions within a small range where the probability of finding it is reasonably good. The bunny is most likely to be found right in the center of the wave packet, but there's a good chance of finding it a little bit to the left, or a little bit to the right. The position as described by this wave packet is necessarily uncertain.

The uncertainty in the wavelength is not as obvious, but it's there because this wave packet is actually a combination of a great many waves, each with a slightly different momentum. Each of these waves represents a particular possible momentum for the bunny, so just as there are several different positions where the bunny might be found, there are also several different possible values of the momentum. The momentum of the bunny described by this wave packet thus has some uncertainty.

How do we get a wave packet by combining many waves? Well, let's start with two simple waves, one corresponding to a bunny casually hopping across the yard, and another one with a shorter wavelength (the graph below shows 20 full oscillations of one, in the same space as 18 of the other), corresponding to a bunny moving faster, perhaps because it knows there's a dog nearby. Now let's add those two wavefunctions together.

"Wait a minute—now we have two bunnies?"

"No, each wavefunction describes a bunny with a particular momentum, but it's the same bunny both times."

"But doesn't adding them together mean that you have two bunnies?"

"No, in this case, it just means that there are two different **states*** you might find the single bunny in. When you look out into the yard, there's some probability of finding the bunny moving slowly, and some probability of finding it moving a little faster. The way we account for that mathematically is by adding the two waves together."

"Oh. Darn. I was hoping for more bunnies."

• • •

*"State" in physics refers to a particular collection of properties—position, momentum, energy, etc. A bunny in the yard with a given momentum is said to be in a momentum state. A second bunny in the same yard with the same momentum is in the same state; a second bunny in the same yard with a different momentum is in a different state. A third bunny with the same momentum as the first, but in a different yard, is in a third state, and so on.

When we add these two waves together, we find that there are some places where they are in phase, and add up to give a bigger wave. In other places, they're out of phase, and cancel each other out. The wavefunction we get from adding them together (the solid line in the figure) has lumps in it—there are places where we see waves, and places where we see nothing. When we square that to get the probability distribution, we get the bottom graph:

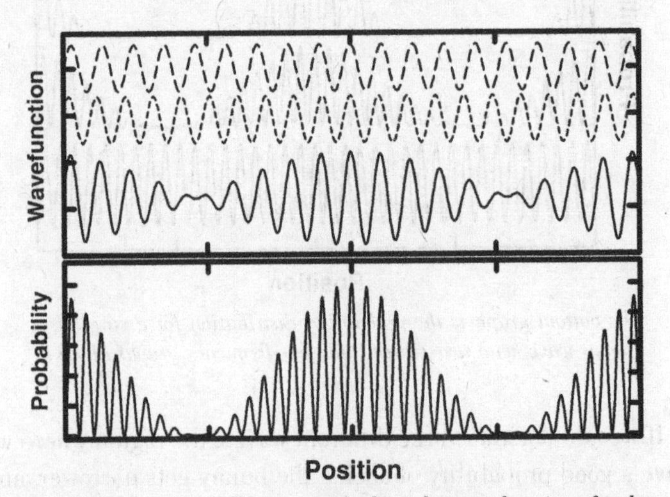

The dashed curves in the top graph show the wavefunctions for the two different wavelengths (shifted up so you can see them clearly). The solid curve shows the sum of the two wavefunctions. The bottom graph shows the probability distribution resulting from adding them together (the square of the solid curve in the top graph).

The center part of this probability distribution looks an awful lot like the wave packet we want. There's a region where we have a good probability of finding the bunny, and in that region, we see a wavelength associated with its motion. Outside that region, the probability goes to zero, meaning that there are places where we have no hope of seeing a bunny at all.

Of course, this two-wave wavefunction isn't exactly what we want, because the no-bunny zone is very narrow and followed immediately by another lump. But we can improve the situation by adding more waves:

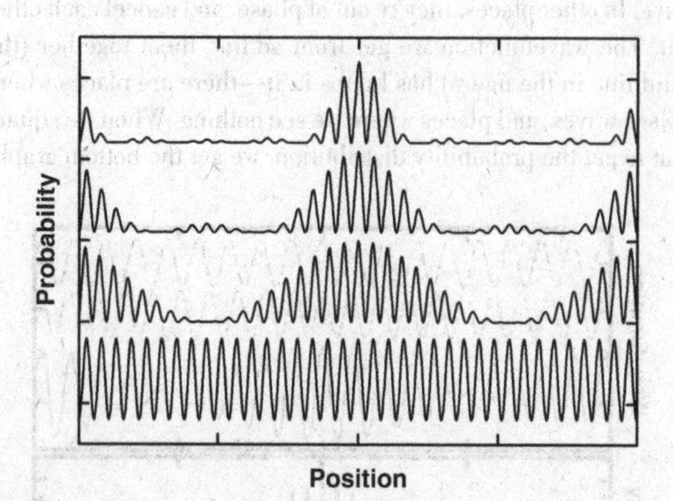

The bottom graph is the probability distribution for a single frequency wave, with two-, three-, and five-frequency graphs above it.

If we add together three different waves, the region where we have a good probability of seeing the bunny gets narrower, and with five different waves, it's narrower still. As we add more and more waves, the regions of high probability get narrower, and the spaces between them become wider and flatter. What we end up with starts to looks like a long chain of wave packets.

"So, now we've got a long chain of bunnies? I thought we were only talking about one bunny."

"We are talking about one bunny. What we've got is a chain of different 'wave packets,' which each describe a different place where we might find the single bunny. If we want to narrow that down to just one position for the bunny, we do it by adding together a continuous distribution of wavelengths, not just a set of regularly spaced single wavelengths."

"Wouldn't that mean adding together an infinite number of different wavelengths, though?"

"Well, yeah, but that's what we have calculus for."

"Oh. I'm not so good at calculus."

"Very few dogs are. Just take my word for it—we can make a single wave packet by summing a continuous distribution of wavelengths, with different probabilities for different wavelengths."

"And we end up with an infinite number of bunnies!"

"Sorry, but no. It's still just one bunny, with an infinite number of possible velocities that are very close to one another."

"Darn. I still want more bunnies."

"Well, the infinite-sum wavefunction does define the position reasonably well, so at least you know where the one bunny is."

"True. And if I know where it is, I can catch it!"

THE LIMITS OF REALITY: THE UNCERTAINTY PRINCIPLE

What does it mean to add together lots of different waves with different wavelengths in this way? Well, each wave corresponds to a particular momentum—a different velocity for the (single) bunny moving through the yard. When we add them all together, what we're doing is saying that there's a chance of finding the bunny in each of those different states (we'll talk more about this in chapter 3).

Adding these states together is the origin of the uncertainty principle. If we want a narrow and well-defined wave packet, so that we know the position of the bunny very well, we need to add together a great many waves to do that. Each wave corresponds to a possible momentum for the bunny, though, which gives a large uncertainty in the momentum—it could be moving at any one of a large number of different speeds.

On the other hand, if we want to know the momentum very

well, we can use a small number of different wavelengths, but this gives us a very broad wave packet, with a large uncertainty in the position. The bunny can only have a few possible speeds, but we can no longer say where it is with much confidence.

We can't produce a wave packet with a single well-defined position without using an infinitely wide distribution of wavelengths, and we can't produce a wave packet with a single well-defined momentum without having it extend over all of space. The best we can hope to do is a single wave packet like we drew at the beginning, with a small uncertainty in the momentum and a small uncertainty in the position. When we go through the mathematical details, we find that the smallest possible product of the uncertainties satisfies the famous Heisenberg relationship*:

$$\Delta x \, \Delta p = h/4\pi$$

The uncertainty in the position (Δx) multiplied by the uncertainty in the momentum (Δp) is equal to Planck's constant divided by 4π. Any other type of wave packet (and there are lots of different shapes found in nature) will have a larger uncertainty product, so the relationship is usually written with a greater-than-or-equal-to sign:

$$\Delta x \, \Delta p \geq h/4\pi$$

*The symbol "Δ" is a capital Greek letter delta, as any fraternity dog can tell you. In science, it's used to indicate a change in something or a difference between two things. Δx is the uncertainty in the position, or the amount of difference you can expect between the position you eventually measure and the most likely position from the wavefunction. When you say "I buried a nice bone sixteen steps from the big oak tree, give or take a step," the "sixteen steps" is the most likely position of the bone, and the "give or take a step" is Δx.

The important result remains the same, though: no matter what you do, there's no way to make both Δx and Δp zero—as you make one smaller, the other necessarily gets larger, and the product remains above the Heisenberg limit.

Looked at in terms of wavefunctions, then, we can see that this relationship is much more than just a practical limit due to our inability to measure a system without disturbing it. Instead, it's a deep statement about the limits of reality. We saw in chapter 1 that quantum particles behave like particles—photons have momentum and collide with electrons in the Compton effect (page 25). We also saw that quantum particles behave like waves—electrons, atoms, and molecules diffract around obstacles and form interference patterns. The price we pay for having both of these sets of properties at the same time is that position and momentum must always be uncertain. The meaning of the uncertainty principle is not just that it's impossible to measure the position and momentum, it's that these quantities do not exist in an absolute sense.

MANIFESTATIONS OF UNCERTAINTY: ZERO-POINT ENERGY

The uncertainty principle forces us to completely rethink our understanding of how the universe works. Not only does it change the way we look at single moving particles, but it has profound consequences for the structure of matter at the microscopic level.

Most humans, and even many dogs, picture atoms as tiny little solar systems, with negatively charged electrons orbiting a positively charged nucleus. This picture originated with Niels Bohr in 1913, when he proposed the first quantum model of the hydrogen atom.

In Bohr's model, the one electron of a hydrogen atom can orbit the nucleus only in certain very specific orbits, with particular well-defined values of energy. These orbits are the "allowed states" of hydrogen, and an electron in an allowed state will happily remain there. Electrons can never be found in an orbit with an in-between energy. Physicists often talk about these states as if they were steps on a staircase, and the electrons were dogs looking for a place to sleep. The dog can rest comfortably on the ground floor, or on one of the steps, but any attempt to lie down halfway between two steps will end badly.

Bohr's model works brilliantly to describe the characteristic colors of light emitted and absorbed by hydrogen. Electrons can move between the allowed states by absorbing or emitting photons of light, with the frequency of the emitted light corresponding to the difference between the energies of the two states. The Bohr model thus solved a problem that had stymied physicists for years.

When Bohr proposed the model, it was a bold break with prior physics. Unfortunately, it's a cobbled-together mix of classical and quantum ideas, with no solid theoretical justification. Louis de Broglie's wave model of the electron provided the missing theoretical basis, but while particle-wave duality justifies the idea of allowed states, it requires us to discard the image of electrons orbiting the nucleus like planets orbiting the sun.

The fundamental problem with this picture is the same issue that leads to uncertainty. For the solar-system model to be accurate, the electron must have a well-defined position somewhere along the allowed orbit, and a well-defined momentum moving it along that orbit. But this can't possibly work—if we try to define the electron's position well enough to locate it along a planetary orbit, it must have a huge uncertainty in momentum, meaning that we can't say where it's going. If we try to define the electron's momentum well enough to place it on an orbital track, it must have a huge uncertainty in position, meaning that we can't even be sure it's near the nucleus it's supposed to be orbiting.

When we account for the wave nature of the electron, we

are forced to discard the whole idea of electrons as planets. Instead, the electron hovers around the nucleus in a fuzzy sort of "cloud," with a position that is uncertain, but confined to a region near the nucleus, and a momentum that is uncertain, but limited to values that keep it near the nucleus. Bohr's idea of allowed energy states still applies—the electron will always have one of the limited number of energy values predicted by Bohr's theory—but these states no longer correspond to electrons moving in particular orbits.

"So, wait—the electron isn't in a particular place, it's just kind of near the atom?"

"That's right. The different energy states correspond to different probabilities of finding the electrons at particular positions, and higher-energy states will give you a better chance of finding the electron farther from the nucleus than lower-energy states. But for any of the allowed states, the electron could be at just about any point within a few nanometers of the nucleus."

"But what happens if you have two atoms close together?"

"Well, if you bring two atoms close enough together, an electron that starts out attached to one atom can end up on the other atom, because of this quantum uncertainty in the position. We'll talk a little more about this in chapter 6, when we talk about tunneling."

"Okay."

"You can also get situations where an electron is sort of 'shared' between two atoms. That's how chemical bonds form. And if you get a bunch of atoms together in a solid, one electron can be shared among the whole solid. That's the basis for the quantum theory of solids, which lets us understand how metals conduct electricity and how to make semiconductor computer chips. It's all because electrons extend beyond specific planetary orbits."

"Uncertain electrons are weird."

"Strictly speaking, it's not just electrons. Everything in the universe is subject to the uncertainty principle, and has an uncertain position and velocity."

"That can't be right. I mean, I can see my bone right over there, and it has a definite position, and a velocity of zero."

"Ah, but the *quantum* uncertainty associated with your bone is dwarfed by the *practical* uncertainty involved in measuring it. If you look at it really carefully, you might be able to specify its position to within a millimeter or so—"

"I *always* look at my bone carefully."

"—and with heroic effort, you might bring that down to a hundred nanometers. In that case, the velocity uncertainty of your hundred-gram bone would be only 10^{-27} m/s. So, the velocity would be zero, plus or minus 10^{-27} m/s."

"That's pretty slow."

"Yeah, you could say that. At that speed, it would take the age of the universe to cross the thickness of a single atom."

"Okay, that's *really* slow."

"We don't see quantum uncertainty associated with everyday objects because they're just too big. We only see uncertainty directly when we look at very small particles confined to very small spaces."

"Like electrons near atoms!"

"Exactly."

Uncertainty has another, even more profound effect on the structure of atoms. Electrons must always have uncertainty in both their position and momentum, and that means that the energy of an electron in an atom can never be zero. To have zero energy while still being part of an atom, an electron would need to be not moving, sitting right on top of the nucleus. This is impossible, as we've already seen—the closest we can come is to make a narrow electron wave packet centered on the nucleus, which will include lots of different states with nonzero momentum. Even the lowest-energy-allowed state of hydrogen, then, has some energy.

This is a general phenomenon, and applies to any confined quantum particle. If we know that a particle is in some particular region of space, that limits the uncertainty in the position, and increases the uncertainty in the momentum. Confined quantum particles are never at rest—they're like puppies in a basket, always squirming and wiggling and shifting around, even when they're asleep.

This tiny residual motion is called **zero-point energy,** which is the minimum quantum energy associated with a particle due to its confinement. Zero-point energy provides an absolute lower limit to the energy a confined particle can have—no matter how carefully you prepare the system, the particles in that system will always be in motion, with small random fluctuations constantly changing the magnitude and direction of their velocity.

Zero-point energy is one of the most counterintuitive ideas in quantum physics, telling us that nothing can ever be perfectly at rest. It means that there is always some energy present in any system, no matter how hard you try to extract all the energy. Even empty space has zero-point energy, which leads to some surprising consequences, including the spontaneous emission of photons from atoms and tiny forces (called "Casimir forces") between metal plates in a vacuum. The zero-point energy of empty space can even produce short-lived pairs of "virtual" particles, as we'll see in chapter 9.

Zero-point energy is probably the most dramatic manifestation of the uncertainty principle. Its existence is a direct consequence of the quantum nature of all the particles making up our universe.

"So, the point of all this is that position and momentum do not have definite values?"

"Yes, that's it exactly."

"Are those the only things this happens with?"

"No, there are lots of different uncertainty relationships between pairs of physical quantities. There's uncertainty in angular momentum, for example—you can't know both the direction of a rotating object and how fast it's spinning. There's also an uncertainty relationship between the number of photons in a light beam and the phase of the wave associated with the beam. Uncertainty relationships are all over the place in quantum physics."

"So, basically, nothing is defined in an absolute sense? Isn't that kind of . . . postmodern?"

"It's not all that bad—it's not like different experimenters get to make up their own results. The uncertainty due to quantum effects is generally very small, so for all practical purposes, we can treat macroscopic objects as if they have definite properties. But at the microscopic level, there really isn't any single defined value for any of these quantities."

"But you talked earlier about measuring bunnies in definite positions. How does that work, if they don't have definite positions?"

"That's a very good question, and takes us into a whole new area of weird stuff, in the next chapter."

CHAPTER 3

Schrödinger's Dog: The Copenhagen Interpretation

I'm in the kitchen, getting a glass of water, when Emmy trots in, tail wagging. "You should give me a treat," she says.

"I should? Why should I give you a treat?"

"Because I'm a very good dog, and I deserve a treat!"

"I'm not going to give you a treat for no reason," I say, "but I'll tell you what I'll do." I reach into the treat jar, then hold out both fists. "Guess where the treat is, and you can have it."

Immediately, her nose starts working.

"No sniffing, either." I put my hands behind my back. "Just guess which hand has the treat."

"Ummm . . . Okay. Both."

"That's not one of the choices."

"But it's the right answer," she says, pouting. "It's like that cat in the box."

"What cat in what box?"

"You know, the one in the box. With the thing. It's dead and alive at the same time. In the box."

"**Schrödinger's cat?**"

"Yeah! That's the one!" She wags her tail excitedly. "I like that experiment. You should do that."

"For one thing, it's just a thought experiment to show the absurdity of quantum predictions. Nobody ever did it for real.

For another, I doubt that people would appreciate it if we started killing cats."

"I don't care about the killing. I just like the idea of putting cats in boxes. Cats belong in boxes."

"I'll pass that on to the scientific community. But what does this have to do with your treat?"

"Well, the treat could be in your left hand, and it could be in your right hand. I don't know which it's in, and you won't let me sniff to see where it is, so that means that the treat is in a **superposition state** of both left and right hands. Until I measure which hand it's in, the answer is that it's in both hands at the same time."

"That's an interesting argument. It doesn't apply here, though."

"Yes it does. It's basic quantum mechanics."

"Well, yeah, it's true that unmeasured objects exist in superposition states as a general matter," I say, "but those superposition states are extremely fragile. Any disturbance at all—absorbing or emitting even a single photon—will cause them to collapse into classical states with a definite value."

"People have seen them, though."

"Sure, there have been lots of 'cat state' experiments done, but the largest superposition anybody has managed to make involved something like a billion electrons.* That's nowhere near the size of a dog treat, which would contain something like 10^{22} atoms."

"Oh."

"And on top of that, even in the most extreme variant of the Copenhagen interpretation, the wavefunction is collapsed by the act of observation by a conscious observer. Now, you can argue about who counts as an observer—"

"Not a cat, that's for sure. Cats are dumb."

*See chapter 4, page 101.

"—but by any reasonable standard, I count as an observer. I know which hand the treat is in. So you're dealing with a classical probability distribution, in which the treat is in either one hand or the other, not a quantum superposition in which the treat is in both hands at the same time."

"Oh. Okay." She looks disappointed.

"So, guess which hand the treat is in."

"Ummm . . . I still say both."

"Why is that?"

"Because I am an *excellent* dog, and I deserve *two* treats!"

"Well, yeah. Also, I'm a sap." I give her both of the treats.

"Ooooo! Treats!" she says, crunching happily.

One of the most vexing things about studying quantum mechanics is how stubbornly classical the world is. Quantum physics features all sorts of marvelous things—particles behaving like waves, objects being in two places at the same time, cats that are both alive and dead—and yet, we don't see any of those things in the world around us. When we look at an everyday object, we see it in a definite classical state—with some particular position, velocity, energy, and so on—and not in any of the strange combinations of states allowed by quantum mechanics. Particles and waves look completely different, dogs can only pass on one side or the other of an obstacle, and cats are stubbornly, irritatingly alive and not happy about being sniffed by strange dogs.

We directly observe the stranger features of quantum mechanics only with a great deal of work, in carefully controlled conditions. Quantum states turn out to be remarkably fragile and easily destroyed, and the reason for this fragility is not immediately obvious. In fact, determining why quantum rules don't seem to apply in the macroscopic world of everyday dogs and cats is a surprisingly difficult problem. Exactly what happens in the transition from the microscopic to the macroscopic has troubled many of the best physicists of the last hundred years, and there's still no clear answer.

In this chapter, we'll lay out the basic principles that are central to understanding quantum physics: **wavefunctions, allowed states, probability**, and **measurement**. We'll introduce a key example system, and talk about a simple experiment that demonstrates all of the essential features of quantum physics. We'll talk about the essential randomness of quantum measurement, and the philosophical problems raised by this randomness, which are disturbing enough that even some of the founders of quantum physics gave up on it entirely.

WHAT DOES A WAVEFUNCTION MEAN? INTERPRETATION OF QUANTUM MECHANICS

Most of the philosophical problems with quantum mechanics center around the "interpretation" of the theory. This is a problem unique to quantum mechanics, as classical physics doesn't require interpretation. In classical physics, you predict the position, velocity, and acceleration of some object, and you know exactly what those quantities mean and how to measure them. There's an immediate and intuitive connection between the theory and the reality that we observe.

Quantum mechanics, on the other hand, is not nearly so obvious. We have the mathematical equations that govern the theory and allow us to calculate wavefunctions and predict their behavior, but just what those wavefunctions *mean* is not immediately clear. We need an "interpretation," an extra layer of explanation, to connect the wavefunctions we calculate to the properties we measure in experiments.

The central elements of quantum mechanics can be presented in many different ways—as many different ways as there are books on the subject—but in the end, they all rest on four basic principles. You can think of these as the core principles of the theory, the basic rules that you have to accept in order to make any progress.*

CENTRAL PRINCIPLES OF QUANTUM MECHANICS

1. **Wavefunctions:** Every object in the universe is described by a quantum wavefunction.
2. **Allowed states:** A quantum object can only be observed in one of a limited number of allowed states.
3. **Probability:** The wavefunction of an object determines the probability of being found in each of the allowed states.
4. **Measurement:** Measuring the state of an object absolutely determines the state of that object.

The first principle is the idea of **wavefunctions**. Every object or system of objects in the universe is described by a wavefunction, a mathematical function that has some value at every point in space. It doesn't matter what you're describing—an electron, a dog treat, a cat in a box—it has a wavefunction, and that wavefunction has some value no matter where you look. The value could be positive, or negative, or zero, or even an imaginary number (like the square root of -1), but it has a value everywhere.

A mathematical formula called the **Schrödinger equation** (after the Austrian physicist and noted cad† Erwin Schrödinger, who discovered it) governs the behavior of wavefunctions. Given

*Sort of like "Thou shalt not climb on the furniture" for dogs living with humans.

†Schrödinger was almost as notorious for his womanizing as for his contributions to physics. He came up with the equation that bears his name while on a ski holiday with one of his many girlfriends, and fathered daughters with three different women, none of them his wife (who, incidentally, knew about his affairs). His unconventional personal life cost him a position at Oxford after he left Germany in 1933, but he carried on living more or less openly with two women (one the wife of a colleague) for many years.

some basic information about the object of interest, you can use the Schrödinger equation to calculate the wavefunction for that object and determine how that wavefunction will change over time, similar to the way you can use Newton's laws to predict the future position of a dog given her current position and velocity. The wavefunction, in turn, determines all the observable properties of the object.

The second principle is the idea of **allowed states**. In quantum theory, an object will only ever be observed in certain states. This principle puts the "quantum" in "quantum mechanics"—the energy in a beam of light comes as a stream of photons, and each photon is one quantum of light that can't be split. You can have one photon, or two, or three, but never one and a half or pi.

Similarly, an electron orbiting the nucleus of an atom can only be found in certain very specific states.* Each of these states has a particular energy, and the electron will always be found with one of those energies, never in an in-between state. Electrons can move between those states by absorbing or emitting light of a particular frequency—the red light of a neon lamp, for example, is due to a transition between two states in neon atoms—but they make those jumps instantaneously, without passing through the intermediate energies. This is the origin of the term "quantum leap" for a dramatic change between two conditions—the actual energy jump is very small, but the change in the state happens in no time at all.

The third principle is the idea of **probability**. The wavefunction of an object determines the probabilities of the different allowed states. If you're interested in the position of a dog, say, the wavefunction will tell you that there's a very good probability of finding the dog in the living room, a lower probability of finding her in the closed bedroom, and an extremely low probability of finding her on one of the moons of Jupiter. If you're inter-

*As we discussed at the end of chapter 2.

ested in the energy of that same dog, the wavefunction will tell you that there's a very good probability of finding her sleeping, a good probability of finding her leaping about and barking, and almost no chance of finding her calmly doing calculus problems.

Philosophical problems start to creep in at this point, because the one thing the wavefunction won't give you is certainty. Quantum theory allows you to calculate only probabilities, not absolute outcomes. You can say that there's some probability of finding the dog in the living room and some probability of finding the dog in the kitchen, but you can't say for sure where she will be until you look. If you repeat the same measurement under the same conditions—asking "Where is the dog?" at four o'clock in the afternoon—you'll get different results on different days, but when you put all the results together, you'll see that they match the probability predicted from the wavefunction. You can't say in advance what will happen for any individual measurement, only what will happen over many repeated experiments.

Quantum randomness is a tremendously disturbing idea for people raised on classical physics, where if you know the starting conditions of your experiment well enough, you can predict the outcome with absolute certainty: you know that the dog will be in the kitchen, and looking just confirms what you already knew. Quantum mechanics doesn't work that way, though: identically prepared experiments can give completely different results, and all you can predict are probabilities. This randomness is the philosophical issue that led Einstein to make a variety of comments that have been rendered as "God does not play dice with the universe."*

*Einstein had many negative things to say about the probabilistic nature of quantum mechanics, but the origin of the usual formulation is a letter to Max Born in 1926, in which he wrote, "The theory delivers a lot, but hardly brings us closer to the secret of the Old One. I for one am convinced that *He* does not throw dice" (quoted in David Lindley's *Uncertainty*, p. 137).

• • •

"Physicists are silly."

"Why do you say that?"

"Well, what's disturbing about randomness? I never know the outcome of anything for sure before it happens, and I'm fine."

"Well, you're a dog, not a physicist. But you do make a good point—any responsible practical treatment of classical physics has to include some element of probability in its predictions, just because you can never account for all the little perturbations that might affect the outcome of an experiment."

"Like that butterfly in Brazil, causing all this weather."

"Exactly. That's the usual metaphor: a butterfly flaps its wings in the Amazon, and a week later, there's a storm in Schenectady. It's the classic example of chaos theory, which shows that probability is unavoidable even in classical physics, because you can never account for every single butterfly that might affect the weather."

"Stupid chaos butterflies."

"The thing is, quantum probability is a different game altogether. The probabilities we end up with in classical physics are a practical limitation. If, by some miracle, you really *could* keep track of every butterfly in the world, then you would be able to predict the weather with certainty, at least for a while. Quantum physics doesn't allow that."

"You mean the butterflies are covered by the uncertainty principle, so you don't know where they are?"

"Only partly—it's deeper than that. In quantum physics, even if you perform the same experiment twice under identical conditions—down to the very last butterfly wing-flap—you still won't be able to predict the exact outcome of the second experiment, only the probability of getting various outcomes. Two identical experiments can and will give you different results."

"Oh. You know what? That is pretty disturbing. Maybe you're not so silly, after all."

"Thanks for the vote of confidence."

The final principle of quantum theory is the idea of **measurement**. In quantum mechanics, measurement is an active process. The act of measuring something creates the reality that we observe.*

To give a concrete example, let's imagine that you have a dog treat in one of two boxes. The boxes are sealed, soundproof (so you can't hear the treat rattling), and airtight (so you can't sniff it out): you can't tell which box the treat is in without opening one of the boxes.

If we want to describe this as a quantum mechanical object, we need to write down a wavefunction with two parts, one part describing the probability of finding the treat in the box on the left, and the other describing the probability of finding the treat in the box on the right. We can do this by adding together the wavefunctions for the treat being in the left-hand box only and the right-hand box only, just as we did in the preceding chapter (page 42) when we made a wave packet by adding together bunny states.

Now, imagine that you open one of the boxes, and find the treat, then close the box back up. You still have one treat and two boxes, but you've measured the position of the treat. What does the wavefunction look like?

The wavefunction now has only one part—the piece describing a treat in the left-hand box—because we know exactly where the treat is. If you found it in the left-hand box, the next time you open that box, there's a 100% chance that it will be there, and

*Werner Heisenberg went so far as to say that the results of measurements were the *only* reality—that it made no sense to talk about where an electron was or what it was doing between measurements.

there's a 0% chance of finding the treat in the right-hand box. The other part that was there before you opened the box, giving the probability of being in the right-hand box, is gone, due to the measurement you made.

Now throw away those boxes, take two new boxes prepared in the same manner as the first pair, and you'll have a two-part wavefunction again. The result of opening the first box won't necessarily be the same, though. You might very well find the treat in the right-hand box this time. If you do, and keep closing and reopening that set of boxes, you'll always find the treat in the right-hand box. Again, you go from a two-part wavefunction to a one-part wavefunction.

So, what's the big deal? After all, that's just how probabilities work, right? In the first experiment, the treat was in the left-hand box all along, but you just didn't know it, and in the second experiment, the treat was in the right-hand box. The state of the treat didn't change, but your knowledge about the state of the treat did.

Quantum probabilities don't work that way. When we have a two-part wavefunction (a "superposition state"), it doesn't mean that the object is in one of the two states, it means that the object is in *both* states *at the same time*. The dog treat isn't in the left-hand box all along, it's simultaneously in both left and right boxes until after you open the box, and find it in one or the other.

"That's pretty strange. Why should we believe it?"

"Well, we can demonstrate the weird features of quantum mechanics with an experiment called a **quantum eraser**."

"Oooh! I like that! Let's erase some cats!"

"It doesn't work on macroscopic objects. It uses polarized light, which I have to explain first."

"Awww . . . Why can't we just erase stuff?"

"I'll keep it as short as I can, but this is important stuff. Polarized light is the best system around for giving concrete examples of quantum effects. We'll need it for this chapter, and also chapters 7 and 8."

"Oh, all right. As long as I can erase stuff later."

"We'll see what we can do."

SUPERPOSITION AND POLARIZATION: AN EXAMPLE SYSTEM

We can show both the existence of superposition states and the effects of measurement using the **polarization** of light. Polarized photons are extremely useful for testing the predictions of quantum mechanics, and will show up again and again in coming chapters, so we need to take a little time to discuss polarization of light, which comes from the idea of light as a wave.

A wave, such as a beam of light, is defined by five properties. We have already talked about four of these: the wavelength (distance between crests in the wave pattern), frequency (how many times the wave oscillates per second at a given point), amplitude (the distance between the top of a crest and the bottom of a trough), and the direction in which the wave moves. The fifth is the polarization, which is basically the direction along which the wave oscillates. An impatient dog owner out for a walk can attempt to get his dog's attention by shaking the leash up and down, which makes a vertically polarized wave in the leash, or by shaking the leash from side to side, which makes a horizontally polarized wave.

Like a shaken leash, a classical light wave has a direction of oscillation associated with it. The oscillation is always at right angles to the direction of motion, but can point in any direction around that (that is, left, right, up, or down, relative to the direction the light is moving). Physicists typically represent the polarization state of a beam of light by an arrow pointing along the direction of oscillation—a vertically polarized beam of light is represented by an arrow pointing up, and a horizontally polarized beam of light is represented by an arrow pointing to the right, as seen in the figure below.

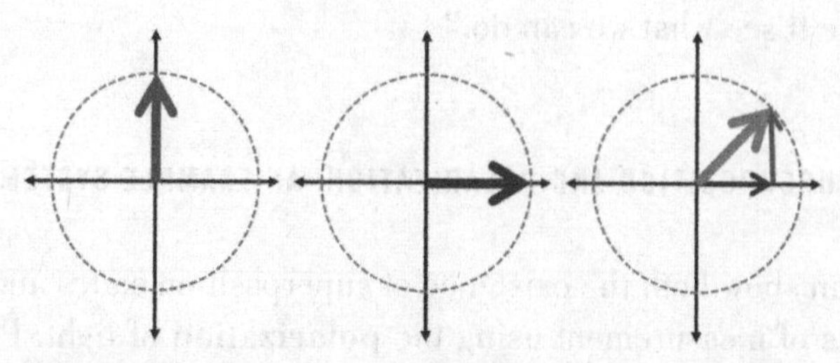

Left: vertical polarization, represented as an up arrow. Middle: horizontal polarization, represented as a right arrow. Right: polarization between vertical and horizontal, which can be thought of as a sum of horizontal and vertical components.

• • •

"Wait, what are these pictures, again?"

"Imagine that you're right behind the beam of light, and looking down the direction of motion. The arrow indicates the direction of the oscillation of the wave. An up arrow means that you'll see the wave moving up and down; a right arrow means that you'll see it moving side to side."

"So . . . an up arrow is like chasing a bunny that bounds up and down, while a right arrow is like chasing a squirrel that zigzags back and forth?"

"Sure, that works."

"Are up and to the right the only options?"

"You can have arrows in other directions, too. An arrow to the left also indicates a side-to-side oscillation, but it's out of phase with the arrow to the right."

"So, a right arrow is a squirrel that zigs to the right first, and a left arrow is a squirrel that zags to the left first?"

"Yeah. If you insist on examples involving prey animals."

"I like prey animals!"

• • •

The polarization of a wave can be horizontal or vertical, but also any angle in between. We can think of the in-between angles as being made up of a horizontal part and a vertical part, as shown in the figure above. Each of these components is less intense (that is, it has a smaller amplitude, indicated in the figure by the length of the arrow) than the total wave, but they add together to give the same final intensity at some angle. You can think of this addition as a combination of steps, just like the way that we can get from one point to another by either taking three steps east followed by four steps north, or by taking five steps in a direction about 37° east of due north.

"So, an in-between angle is like a bunny that's zigzagging left and right, while also hopping up and down?"

"Yes, that's right."

"Or a squirrel that's jumping up and down while it zigzags left and right?"

"I think that's about enough prey examples for now."

"You're no fun."

Thinking of in-between polarizations as a sum of horizontal and vertical components is a useful trick because it makes it easy to see what happens when light encounters a **polarizing filter**. Polarizing filters are devices that will allow light polarized at a particular angle—vertical, say—to pass through unimpeded, while light polarized at an angle 90° away—horizontal—will be completely absorbed. You can understand the effect by imagining a dog on a leash that passes through a picket fence. If you shake the leash up and down, the wave will pass right through, but side-to-side shaking will be blocked by the boards of the fence.

When light at an angle between vertical and horizontal strikes a vertically oriented polarizing filter, only the vertical component of the light will pass through. This lowers the intensity of the light on the other side, by an amount that depends on the angle. For small angles, most of the light makes it through—at an angle of 30°, the beam on the far side is three-fourths as bright as the initial beam—while for larger angles, most of the beam is blocked—at 60° from vertical, the beam on the far side is only one-fourth as bright as the initial beam. At an angle of 45°, midway between horizontal and vertical, exactly half of the light will pass through the filter.

The light on the far side of the filter is polarized at the angle of the filter, no matter what angle it started at. For this reason, polarizing filters are commonly called **polarizers:** light passing through a vertically oriented polarizing filter will emerge as vertically polarized light, whether it started with vertical polarization or at some other angle. The overall amount of light will be different, but the polarization will be the same. All of the light passing through a vertically oriented filter will pass through a second vertical filter, and all of it will be blocked by a horizontally oriented filter.

"What is all this good for, anyway?"

"Other than helping demonstrate quantum physics? Plenty. Light polarization is an extremely useful thing. Digital displays on watches, cell phones, and televisions use a polarizer in front of a light source to vary the amount of light that gets through. And polarizing filters are also used to make sunglasses."

"Sunglasses?"

"Yeah, those sunglasses that I wear when I take you for walks are actually polarizing filters. The light from the sun is unpolarized—it's as likely to be horizontal as vertical—but when light reflects off a surface, it tends to become slightly polarized. Light reflecting off the road out in front of us when we're walking has more horizontal polarization than vertical, so by wearing vertical polarizers as sunglasses, I can block most of that light."

"What's the point of that? Doesn't it make it harder to see?"

"Actually, it reduces the glare off the road, and makes it easier to see things up ahead."

"Things . . . Like bunnies in the road?"

"For example, yes."

"Can I have some polarized sunglasses so I can see bunnies?"

"The ones I have won't fit on your ears, but we'll look into it. Later. First, I have to talk about quantum measurement with polarized light."

"Oh, yeah. Quantum physics. Right."

How does all this apply to light as a particle, though? We spent a good chunk of chapter 1 describing how a beam of light is both a stream of photons and a smooth wave. The last few pages have been discussing polarization in classical terms. How do we handle light polarization in quantum physics?

When we're dealing with classical light waves, it's easy to understand how part of a wave can pass through the filter. When we talk about light in terms of photons, though, the filter is an all-or-nothing proposition. Any given photon either makes it through, or gets absorbed by the filter. There are no "parts" of photons.

We handle the interaction between photons and polarizing filters by saying that each photon has a probability of passing through the filter that is equal to the fraction of the total wave that makes it through in the classical model. If a beam of light with a polarization at 60° from vertical encounters a vertical polarizing filter, the beam on the far side will be one-fourth as bright, meaning that it has one-fourth as many photons. That means that each individual photon has only one chance in four of making it through the polarizing filter.

Each photon making it through the filter will also have its polarization determined by the filter. Only one photon in four may make it through a vertically oriented polarizing filter, but every one of those photons will pass through a second vertical filter, and none of them will pass through a horizontal filter.

"Vertical" and "horizontal" are then the allowed states of the single photon's polarization—when we measure the polarization using a filter, we will find the photon in one of those two states (either passing through the vertical filter, or being absorbed by it), and not anywhere in between.

Polarized photons thus provide an excellent system for looking at the core principles of quantum mechanics. Each individual photon can be described in terms of a **wavefunction**, with two parts corresponding to the two **allowed states**, horizontal and vertical polarization. That wavefunction gives you the **probability** of the photon passing through a polarizing filter, and after you make a **measurement** of the polarization with a filter, the photons are in only one of the allowed states. A single photon passing through a polarizing filter demonstrates all the essential features of quantum physics. As a result, polarized photons have been used in many experiments demonstrating quantum phenomena.

"So, let me get this straight. A photon at an angle between horizontal and vertical is in a superposition state? And sending it through a polarizing filter is the same as measuring it?"

"Yes. You get all the features of quantum superposition and measurement—wavefunctions, allowed states, probability, and measurement—using single polarized photons."

"But I thought you said all this stuff worked the same way when you talked about light as a classical wave?"

"Well, yeah. The end result is the same as the classical polarized wave description."

"What's the big deal, then? I mean, your big example of quantum weirdness is something that just reproduces classical results?"

"Well, no. I mean, that's not my big example. The big example of quantum weirdness is in the next section."

"Oh. Well, carry on, then."

(UN)MEASURING A PHOTON: THE QUANTUM ERASER

One of the best demonstrations of the weirdness of quantum superpositions is an experiment called a **quantum eraser.** The quantum eraser encapsulates everything that's strange about single-particle quantum physics in a single experiment: particle-wave duality, superposition states, and the active nature of measurement. If you can understand the quantum eraser, you've understood the essential elements of quantum physics.

Many different variants of quantum-eraser experiments have been done over the years,* but the simplest starts with a variant of Young's double-slit experiment (page 18). If we send a beam of photons at a pair of narrow slits, we will see an interference pattern on the far side of the slits, built up out of single photons detected at particular points (as shown in the figure on the next page). We can see the pattern only because light passes through both slits at the same time. If we block one slit, the interference pattern will disappear, and we'll see only a broad scattering of photons due to the light passing through the unblocked slit.

The interference pattern that we see indicates that the photons are in a superposition state: the wavefunction describing each photon has two parts, one for the photon passing through the left slit, and the other for the photon passing through the right slit. Each photon has passed through *both* slits, at the same time, and the interference between those two components is what produces the pattern we see. Interference patterns always turn up when you have a two-part wavefunction. When we block one slit, we only have a one-part wavefunction, destroying the superposition, and there is no interference pattern.

*The April 2007 issue of *Scientific American* even describes a quantum-eraser experiment that you can do at home, using a laser pointer, tinfoil, wire, and a few pieces of cheap polarizing film.

Interference pattern built up from single photons. Left to right, 1/30 second, 1 second, 100 seconds. Images by Lyman Page at Princteon University, reprinted with permission.

• • •

"Wait, I thought the interference was between two different photons—one that went through the left slit, and one that went through the right slit?"

"That's an easy thing to think, since we usually send in lots of photons at the same time. We can show that that's not the case, though, by sending light at the slits one photon at a time."

"How does a single photon give you an interference pattern?"

"It doesn't. Each individual photon shows up as a single spot, at a particular position on the screen, and where any individual photon turns up is random."

"There's the probability thing again."

"Exactly. The individual photons are random, but if you repeat the experiment over and over again, and keep track of all the photons, you'll see them add up to form an interference pattern. There are some places where you're very likely to find a photon, and other places where there's absolutely no chance of finding one. The overall pattern is determined by the probability distribution you get from the wavefunction for each individual photon interfering with itself."

"So it's one particle, but it goes through both slits, and then ends up at one place on the other side?"

"Exactly."

"That's just weird."

"That's quantum physics."

Instead of blocking one slit, though, let's imagine covering the two slits with two different polarizing filters, one vertical and one horizontal. We put a filter on the left slit that will pass only horizontally polarized light, and we put a filter on the right slit that will pass only vertically polarized light. If we send in light polarized at an angle of 45° to the vertical, it has a 50% chance of going through a horizontal polarizing filter, and a 50% chance of going through a vertical polarizing filter, so we get some light through each slit.

This arrangement of filters gives us a way of measuring which slit the light went through. If we put a vertical polarizer in front of our detector, we will only see light that went through the right-hand slit, and if we put a horizontal polarizer in front of our detector, we will only see light that went through the left-hand slit. The polarizer in front of the detector lets us tell which slit the photon went through, just as if we had put a detector right next to the slit and measured the position directly.

What happens when we do this? When we look at the light with the filters over the two slits, we don't see any sign of an interference pattern. When we measure the polarization of the light, we measure which slit the light went through, and that takes us from a two-part wavefunction, which produces an interference pattern, to a one-part wavefunction, which does not. The act of measuring which slit the photon went through destroys the component of the wavefunction describing the photon going through the other slit, just as the act of opening one of the boxes destroyed the component for the treat being in the other box.

We don't even need to put a polarizing filter on the detector—by putting the polarizers over the slits, we have "tagged" each photon, and the mere fact that we *can* measure which slit it went through is enough to destroy the pattern. In the treats-in-boxes example, this is like someone writing "Treat" on the outside of the box containing the treat—we no longer need to open the box to destroy the superposition.

The disappearing pattern is pretty weird in its own right, but things get weirder: we can undo the measurement after the fact by using a 45° polarizer instead of a horizontal or vertical polarizer to look at the light after the slits. If we do this, we see an interference pattern again! A 45° polarizer will pass either horizontal or vertical polarization, each with a 50% probability, which means any light we detect after the polarizer could have gone through either of the two slits, or even both at once. The third filter "erases" the information we had gained by tagging the photon, like somebody removing the label from the box containing the treat. Inserting the extra polarizer makes it as if we had never made the measurement at all. The second part of the wavefunction isn't destroyed after all, and we can see interference.

The quantum-eraser experiment encapsulates everything that is strange about the core principles of quantum mechanics. The appearance of the interference pattern shows the superposition of quantum states, as each photon goes through both slits at the same time, and the disappearance and reappearance of the pattern when we add polarizing filters shows the active nature of quantum measurement. Just the fact that it is possible to measure which slit the particle went through is enough to completely change the results of the experiment.

WHAT YOU SEE IS ALL THERE IS: THE COPENHAGEN INTERPRETATION

These four ideas—**wavefunctions, allowed states, probability**, and **measurement**—are the central elements of quantum theory. The interference patterns we see in experiments with photons* confirm that quantum particles really do occupy multiple states at the same time. The disappearance and reappearance of the pattern in the quantum-eraser experiment confirms that measurement is an active process and determines what happens in subsequent experiments.

We still have a problem, though, because there is no mathematical process for describing how to get from a probability to the result of a measurement. We use the Schrödinger equation to calculate the wavefunctions for the allowed states of a physical object, and we use the wavefunction to calculate the probability distribution, but we cannot use the probability distribution to predict the exact result of an individual measurement. Something mysterious happens in the process of making a measurement.

This "measurement problem" is the origin of the competing interpretations of quantum mechanics, and the point where physics is forced to become philosophy. All interpretations use the same methods to calculate probabilities for the outcomes of repeated measurements. They differ only in how they explain the step from a quantum superposition state, where the wavefunction consists of two (or more) states at the same time, to the classical result of a single measurement, where the object is found in one and only one state.

The first interpretation put forward for quantum theory was developed by Niels Bohr and coworkers at his institute in Den-

*And electrons, and atoms, and molecules . . .

mark, and is thus known as the **Copenhagen interpretation.** The Copenhagen interpretation is a very ad hoc approach to the problem of measurement in quantum mechanics (which is in some ways typical of Bohr's approach*).

The Copenhagen interpretation tries to avoid the problems of superposition and measurement by drawing a strict line between microscopic and macroscopic physics. Microscopic objects—photons, electrons, atoms, and molecules—are governed by the rules of quantum mechanics, but macroscopic objects—dogs, physicists, and measurement apparatus—are governed by classical physics. There's an absolute separation between the two, and you will never see a macroscopic object behaving in a quantum manner.

Quantum measurement involves the interaction of a macroscopic measurement apparatus with a microscopic object, and that interaction changes the state of the microscopic object. The usual description is that the wavefunction "collapses" into a single state. This "collapse," in the Copenhagen interpretation, is an actual change of the wavefunction from a spread-out quantum state with multiple possible measurement outcomes to a state with a single measured value.†

In the most extreme variants of the Copenhagen interpretation, the collapse requires not only a macroscopic measurement

*As we said last chapter (page 49), Bohr's first great contribution to physics was a simple quantum model of hydrogen. It was a cobbled-together mix of quantum and classical ideas with no clear justification that happened to give the right result, and it's unclear what led Bohr to put it forth. It did, however, point the way toward the modern quantum theory that we're discussing in this book.

†The word "collapse" has come to be strongly associated with Copenhagen-type interpretations. There are other approaches to the problem of the projection of a multicomponent wavefunction onto a single measurement result that don't involve a physical change in the wavefunction. We'll look at the best-known example of these "no-collapse" interpretations in chapter 4.

apparatus, but also a conscious observer to note the measurement. In this view, a tree that falls in the forest hasn't really fallen until some person (or dog) comes along and observes it.

"So, wait, doesn't that mean rejecting the entire idea of an objective physical reality?"

"In its most extreme forms, yes. Werner Heisenberg was probably the most radical of the Copenhagen crowd, and he insisted quite strongly that it was a mistake to talk about electrons having an independent reality. In Heisenberg's view, the only things we can really talk about are the outcomes of specific measurements. He rejected the whole notion of talking about what the electrons were doing between measurements."

"That's . . . pretty radical. I don't think I like that."

"You're not alone, believe me."

The Copenhagen interpretation raises a great many problems, among them that there's no obvious reason for the absolute distinction between microscopic and macroscopic physics. As we've already seen, while it gets more difficult to detect quantum behavior as objects get larger and more complicated, it is still possible to see wave behavior in rather large molecules. Macroscopic objects *ought* to be described by quantum wavefunctions and quantum rules.

Another problem is who or what counts as an "observer" for the purposes of collapsing the wavefunction. The requirement that somebody observe the outcome of a measurement before the measurement really "counts" seems to assign some sort of mystical quality to "consciousness," and that idea makes many physicists uncomfortable.

Even the idea of the "collapse" itself is problematic. No mathematical formula exists to describe the collapse—you can use the Schrödinger equation to describe how a wavefunction changes between measurements, but there is no way to describe the "collapse" process. All you can do is choose a result, and start over with a new wavefunction after the measurement. Many physicists find this a little too magical for comfort.

The most famous illustrations of the problems with the Copenhagen interpretation are the infamous "Schrödinger's cat" thought experiment, and the follow-up thought experiment of "Wigner's friend." Despite his role in creating quantum theory, Erwin Schrödinger, like Einstein, had deep philosophical problems with its interpretation, and became disillusioned with the entire field. Schrödinger's cat, which is arguably more famous than his equation, is a diabolical thought experiment through which Schrödinger attempted to illustrate the absurdity of the Copenhagen interpretation. He imagined placing a cat in a sealed box with a radioactive atom that has a 50% chance of decaying within one hour, and a device that will release poison gas if the atom decays, killing the cat. What, he asked, is the state of the cat at the end of the hour?

As Schrödinger noted, according to the Copenhagen interpretation the wavefunction describing the cat would be equal parts "alive" and "dead." This would last until the experimenter opens the box, at which point it would collapse into one of the two states.* This seems completely absurd, though—the idea of a cat that is both dead and alive at the same time is outlandish. And yet this is exactly what seems to happen with photons.

The Copenhagen interpretation also seems to be saying that physical reality does not exist until a measurement is made, which poses its own philosophical problems. Eugene Wigner

*Or, as the British writer Terry Pratchett described it in his novel *Lords and Ladies*, applied to a particularly nasty cat: "Technically, a cat locked in a box may be alive or it may be dead. You never know until you look. In fact, the mere act of opening the box will determine the state of the cat, although in this case there were three determinate states the cat could be in: these being Alive, Dead, and Bloody Furious" (p. 226, Harper paperback).

brought this out by adding another layer to the cat experiment, imagining that the entire thing was conducted by a friend, and only reported to him later. Wigner asked when the wavefunction collapsed: When the friend opened the box, or later, when Wigner heard the result? Has a tree in a forest really fallen before your dog tells you that it's on the ground?

None of the Copenhagen interpretation's answers to these questions are very satisfying, philosophically. While quantum mechanics does an outstanding job of describing the behavior of microscopic objects and collections of objects, the world we see remains stubbornly, infuriatingly classical. Something mysterious happens in the transition from the weird world of simple quantum objects to the much larger world of everyday objects. The Copenhagen approach of insisting on an absolute division between microscopic and macroscopic strikes many physicists as simply dodging the question: it says *what* happens, but not *why*.

How best to handle the transition between quantum and classical remains a subject of active debate. Some future theory may lead to a detailed understanding of what, exactly, happens when we make a measurement of a quantum object. Until then, we're stuck with one of the various interpretations of quantum mechanics.

"I don't think I like this interpretation. It's awfully solipsistic, isn't it?"

"You're not alone. There aren't very many physicists these days who are really happy with the Copenhagen interpretation."

"So, what interpretation do you like?"

"Me? I tend to go with the 'shut up and calculate' interpretation. The name is sometimes attributed to Richard Feynman,*

*Feynman tends to get credit for anything clever said by a physicist in the latter half of the twentieth century. "Shut up and calculate" probably isn't

but the idea is just to avoid thinking about it. Quantum mechanics gives us very good tools for calculating the results of experiments, and the question of what goes on during measurement is probably better left to philosophy."

"I don't think I like that one, either. It's hard to work a calculator without opposable thumbs."

"Well, there are all sorts of different interpretations—there's the many-worlds interpretation, David Bohm's nonlocal mechanics, and something called the 'transactional interpretation.' There are almost as many interpretations of quantum mechanics as there are people who have thought deeply about quantum mechanics."

"I like the many-worlds interpretation. You should talk about that."

"Good idea. That's the next chapter."

"I knew that."

Feynman, though—its first appearance in print seems to be a David Mermin column in *Physics Today* (April 1989, p. 4), as he explains in the May 2004 issue (p. 10).

CHAPTER 4

Many Worlds, Many Treats: The Many-Worlds Interpretation

I'm sitting at the computer typing, when Emmy bumps up against my legs. I look down, and she's sniffing the floor around my feet intently.

"What are you doing down there?"

"I'm looking for steak!" she says, wagging her tail hopefully.

"I'm pretty certain that there's no steak down there," I say. "I've never eaten steak at the computer, and I've certainly never dropped any on the floor."

"You did in some universe," she says, still sniffing.

I sigh. "All right, what ridiculous theory has your silly little doggy brain come up with now?"

"Well, it's possible that you would eat steak at the computer, yes?"

"I do eat steak, yes, and I sometimes eat at the computer, so sure."

"And if you were to eat steak at the computer, you'd probably drop some on the floor."

"I don't know about that . . ."

"Dude, I've seen you eat." Yes, the dog calls me "dude." There may be obedience classes in her future.

"All right, we'll allow the possibility."

"Therefore, it's possible that you dropped steak on the floor.

And according to Everett's **many-worlds interpretation** of quantum mechanics, that means that you *did* drop steak on the floor. Which means I just need to find it."

"Well, technically, what the many-worlds interpretation says is that there's some branch of the unitarily evolving wavefunction of the universe in which I dropped steak on the floor."

"Ummm . . . yeah. Right. Anyway, I just need to find the unitary whatsis."

"The thing is, though, we can only perceive one branch of the wavefunction."

"Maybe *you* can only perceive one branch. I have a very good nose. I can sniff into extra dimensions. They're full of evil squirrels. With goatees."

"That's *Star Trek*, not science, and anyway, extra dimensions are a completely different thing. In the many-worlds interpretation, once there has been sufficient decoherence between the branches of the wavefunction that there's no possibility of interference between the different parts, they're effectively separate and inaccessible universes."

"What do you mean, decoherence?"

"Well, say I did have a piece of steak here—stop wagging your tail, it's hypothetical—quantum mechanics says that if I dropped it on the floor, then picked it back up, there could be an interference between the wavefunction describing the bit of steak that fell and the wavefunction describing the bit of steak that didn't fall. Because, of course, there's only a probability that I'd drop it, so you need both bits."

"What would that mean?"

"I'm not really sure what that would look like. The point is, though, it doesn't really matter. The steak is constantly interacting with its environment—the air, the desk, the floor—"

"The dog!"

"Whatever. Those interactions are essentially random, and

unmeasured. They lead to shifts in the wavefunctions of the different bits of steak, and those shifts make it so the wavefunctions don't interfere cleanly anymore. That process is called **decoherence**, and it happens very fast."

"How fast?" she asks, looking hopeful.

"It depends on the exact situation, but as a rough guess, probably 10^{-30} seconds or less."

"Oh." She deflates a little. "That's fast."

"Yeah. And once that decoherence has happened, the different branches of the wavefunction can't interact with one another anymore. Which means, essentially, that the different branches become separate universes that are completely inaccessible to one another. Things that happen in these other 'universes' have absolutely no effect on what happens in our universe."

"Why do we only see one branch of the whatchamacallit?"

"Ah, now that's the big question. Nobody really knows. Some people think this means that quantum mechanics is fundamentally incomplete, and there's a whole community of scientists doing research into the foundations of quantum theory and its interpretations. The important thing is, there's no way you're going to find steak under my desk in this universe, so please get out of there."

"Oh. Okay." She mopes out from under the desk, head down and tail drooping.

"Hey, look on the bright side," I say. "In the universe where a version of me dropped a piece of steak on the floor, there's also a version of you."

"Yeah?" Her head picks up.

"Yeah. And you're a mighty hunter, so you probably got to the steak before I could pick it up."

"Yeah?" Her tail starts wagging.

"Yeah. So, in the universe where I dropped steak, you got to eat steak."

"Oooh!" The tail wags furiously. "I like steak!"

"I know you do." I save what I was working on. "Tell you what, how about we go for a walk?"

"Ooooh! Good plan!" And she's off, clattering down the stairs for the back door and the leash.

Few physicists have ever been entirely happy with the Copenhagen interpretation discussed in the previous chapter. Numerous alternatives have been proposed, each attempting to find a more satisfying way to deal with the problem of quantum measurement. The most famous of these is commonly known as the **many-worlds interpretation**, which has achieved a dominant position in pop culture, if not among physicists, thanks to its prediction of a nearly infinite number of alternate universes in which events took a different path than the one we see. It's a wonderful science fiction conceit, turning up in books, movies, and the famous *Star Trek* episode featuring an evil Spock with a goatee.

In this chapter, we'll talk about the many-worlds interpretation, and how it addresses some of the problems raised by the Copenhagen interpretation. We'll also discuss the physical process known as "decoherence," in which fluctuating interactions with the environment obscure the effects of interference between different parts of the wavefunction. Decoherence is central to the modern understanding of quantum mechanics, and may be the critical factor for understanding the move from the microscopic world of quantum physics to the classical world of everyday objects.

THEN A MEASUREMENT OCCURS: PROBLEMS WITH COPENHAGEN

The most disturbing element of the Copenhagen interpretation by far, for a physicist at least, is the lack of a mathematical procedure for describing what happens when you make a measurement of some quantity. The Schrödinger equation allows you to calculate what happens to the wavefunction between measurements, but at the instant of a measurement the Copenhagen interpretation says that normal physics stops, and something happens to select a single outcome in a way that does not involve any known mathematical equation.

The ad hoc nature of the Copenhagen interpretation, with its arbitrary division between microscopic and macroscopic physics and its mysterious "wavefunction collapse," is tremendously disturbing, because the whole project of modern theoretical physics is to find a single consistent mathematical description of the world. The unexplained process of wavefunction collapse is like the famous Sidney Harris cartoon of a scientist who has written "Then a miracle occurs" as the second step of a problem. Normal science has no room for miracles, and the Copenhagen collapse idea is a little too miraculous for comfort.

Most physicists (particularly experimentalists) are content to use the idea of wavefunction collapse as a calculational shortcut and go about the business of predicting and measuring the physical world, for which regular quantum theory works astoundingly well. In this "shut up and calculate" interpretation, the problem of finding a consistent explanation for quantum measurement is pushed aside to be dealt with by philosophers. Some better theory may eventually come along, but until then, we should do what we can with what we've got (which turns out to be an awful lot).

The nature of measurement has been a problem from the first days of quantum theory, though, and a few physicists have always chosen to think deeply about these issues. Many of these physicists think that the lack of a clear explanation for the "collapse" of the wavefunction indicates that the Copenhagen interpretation is fundamentally flawed. Thus, they have always searched for some alternative interpretation.

THERE IS NO COLLAPSE: HUGH EVERETT'S MANY-WORLDS INTERPRETATION

In 1957, a graduate student at Princeton named Hugh Everett III suggested a solution to the "collapse" problem that's breathtaking in its simplicity. The reason there is no mathematical method to describe the collapse of the wavefunction, Everett said, is because there is no such thing as the collapse of the wavefunction. The wavefunction always and everywhere evolves according to the Schrödinger equation, but we only see a small piece of the larger wavefunction of the universe.

Let's return to the previous chapter's example of dog treats in sealed boxes to see how this works. If we imagine that we have one dog treat in two boxes, the Copenhagen picture says that we initially have a wavefunction for the treat that consists of two pieces at the same time. This wavefunction changes in time according to the Schrödinger equation. When we open a box and let the dog look inside, the wavefunction instantaneously collapses into only one of those two states, with the treat in either the left-hand box or the right-hand box. We predict future changes by starting over with the Schrödinger equation using the new one-part wavefunction.

In the Everett picture, there is no collapse. The wavefunction starts out in a superposition, a two-part wavefunction with pieces corresponding to the treat in both left-hand and right-hand boxes, and when we open a box that superposition just becomes a little bigger. Now the superposition includes not just the apparatus but also the dog measuring the position of the treat. One piece is "treat in the left-hand box plus a dog who knows the treat is in the left-hand box," and the other is "treat in the right-hand box plus a dog who knows the treat is in the right-hand box." This process continues as you move into the future. If the next step in the experiment involves the dog either eating the treat or not (a low probability outcome, but it's possible), the wavefunction contains four pieces: a dog who ate the treat from the left-hand box; a dog who didn't eat the treat from the left-hand box; a dog who didn't eat the treat from the right-hand box, and a dog who ate the treat from the right-hand box.

The increase in complexity is even more striking in mathematical notation. We start with a two-component wavefunction for just the treat:

$$\Psi_{total} = |L>_{treat} + |R>_{treat}$$

where the angled brackets represent wavefunctions for the treat being in the left or right boxes. Then we bring in the dog:

$$\Psi_{total} = |L>_{treat}|L>_{dog} + |R>_{treat}|R>_{dog}$$

and finally, the decision to eat the treat or not:

$$\Psi_{total} = |L>_{treat}|L>_{dog}|eat> + |L>_{treat}|L>_{dog}|not>$$
$$+ |R>_{treat}|R>_{dog}|eat> + |R>_{treat}|R>_{dog}|not>$$

As you can see, this gets very complicated very quickly, but its evolution is always described by the Schrödinger equation.

"You know, I'm not getting a lot out of these equations."

"You're not supposed to understand them in detail. They're just there to illustrate the increasing complexity of the wavefunction in a more compact manner."

"So, basically, they're just supposed to look scary?"

"Pretty much."

"Oh. Good job, then."

• • •

The Everett picture doesn't immediately appear to be an improvement. The mysterious "collapse" is removed, but at the price of a wavefunction that's expanding exponentially. At first glance, it also seems to defy reality, as we never see systems in more than one state. If all these extra pieces of the wavefunction are running around, why don't we perceive objects as being in multiple states at the same time?

The answer, according to Everett, is that we can't separate the observer from the wavefunction. The observer is included with the rest of the system—the components are things like "treat in the left-hand box plus a dog who knows the treat is in the left-hand box"—and as a consequence, we only perceive our own small part of the overall wavefunction. This branching is the origin of the quantum randomness that Einstein and others found so troubling: the wavefunction always evolves in a smooth and continuous way, but we only experience one branch of the wavefunction at a time, and which branch we see is a random choice. Other versions of ourselves exist in the other branches, experiencing different outcomes (for this reason, the interpretation is sometimes referred to as the "many-minds" interpretation).

None of the myriad other branches have any detectable influence over the events in our branch, and our branch has no detectable influence over the events in any of the others. For all intents and purposes, those other branches are self-contained parallel universes, completely inaccessible from our universe. This is the origin of the name "many-worlds" for the theory: it's as if the universe forks every time a measurement is made, and is constantly spawning new parallel universes with slightly different histories.

WAVEFUNCTIONS FALL APART: DECOHERENCE

These noninteracting branches present a serious but subtle problem for the many-worlds interpretation. Every other two-part wavefunction we've seen has led to some sort of interference phenomenon. So, why don't we see interference around us all the time if there are all these extra branches to the wavefunction? What is it that seals these "parallel universes" off from us?

The answer is a process called **decoherence**, which prevents the different branches of the wavefunction from interacting with one another. Decoherence is the result of random, fluctuating interactions with a larger environment, which destroy the possibility of interference between different branches of the wavefunction, and make the world we experience look classical. Decoherence doesn't just occur in the many-worlds interpretation of quantum mechanics—it's a real physical process, compatible with any interpretation*—but it's particularly important in the modern view of many-worlds (which is sometimes called "decoherent histories" as a result—the interpretation has almost as many names as universes†).

Decoherence is absolutely critical to the modern view of quantum mechanics and quantum interpretations. The most common semiclassical explanations of decoherence leave a lot to be desired, though, as they are inaccurate, and often somewhat circular. The real theory of decoherence is subtle and difficult to understand. As with the uncertainty principle (chapter 2, page

*In fact, all of the viable candidates to replace the Copenhagen interpretation include decoherence as an important part of the measurement process.

†"Many-worlds," "many-histories," "many-minds," "decoherent histories," "relative state formulation," and "theory of the universal wavefunction," among others.

40), though, it's worth some effort to unpack it, because it provides a much richer understanding of the way the universe works.

To understand the idea of decoherence, let's think about the concrete example of a simple interferometer, a device consisting of two beam splitters that split a beam of light in half and a couple of mirrors that bring it back together again.* Interferometers like this are extremely important in physics, not only for demonstrating quantum effects, but because they form the basis for the world's most sensitive detectors of rotation, acceleration, and gravity. These allow the measurement of tiny forces in physics experiments, and also find application in submarine navigation.

Light enters the interferometer when it strikes a beam splitter, which passes half of the light through without perturbing it and reflects the other half off at a 90° angle. These two beams separate from each other, and then are steered back together using two mirrors, and recombined on a second beam splitter. The second beam splitter is lined up so that the transmitted light from one beam and the reflected light from the other follow exactly the same path and interfere with each other before falling on one of two detectors.

You might think that each of the two detectors would detect exactly half of the light, because each detector receives one-quarter of the original beam from each of the two paths (¼ + ¼ = ½). Each detector can actually see anything from no light at all up to the full intensity of the initial beam, though, because the waves that took different paths interfere with each other, like the waves in the double-slit experiment in chapter 1 (page 18).

If the paths followed by the two beams have exactly the same length, the two light waves undergo the same number of oscillations en route to Detector 2, and interfere constructively, giv-

*This is called a "Mach-Zehnder interferometer" after the German and Swiss physicists who invented it.

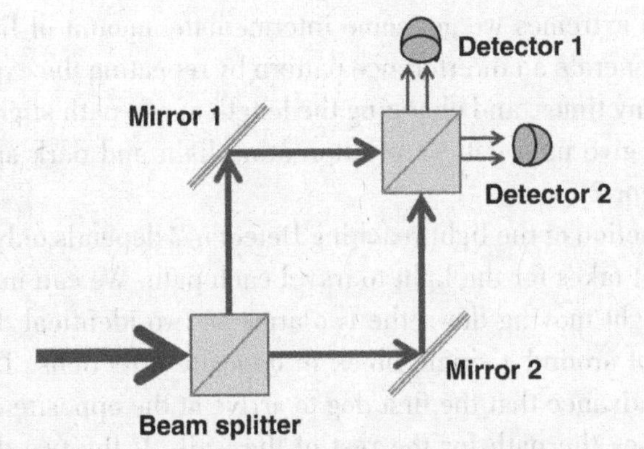

The interferometer described in the text. Light enters from the left, is split by a beam splitter into two different paths, and then recombined by the second beam splitter. If the beam hitting Mirror 1 travels exactly the same distance as the beam hitting Mirror 2, interference causes all the light to hit the upper detector (Detector 1), and none to hit the detector at right (Detector 2).

ing a bright spot—all of the light entering the interferometer hits that detector.* On the other hand, if one path is longer than the other by one half of the wavelength of the light, the light along that path undergoes an extra half-oscillation, and the two waves interfere destructively: the crests of the wave hitting Mirror 1 line up with the troughs of the wave hitting Mirror 2, and they cancel out, giving no light at Detector 2. If we increase the length difference to one full wavelength, the peaks align again, and we get another bright spot, and so on. Between

*There is a small shift in the "phase" of a wave when it reflects off a beam splitter, as if the reflected beam had traveled a small extra distance. As a result, when the two paths have equal length, the waves hitting Detector 2 are in phase, and add to give a bright spot, while the waves reaching Detector 1 are out of phase, and cancel each other out. For paths with different lengths, the two detectors give complementary signals—when one sees no light, the other detects the full initial intensity, and vice versa.

those two extremes we get some intermediate amount of light. We can generate an interference pattern by repeating the experiment many times, and changing the length of one path slightly. This will give us a pattern of alternating light and dark spots on Detector 2.

The fraction of the light reaching Detector 2 depends only on the time it takes for the light to travel each path. We can imagine the light moving down the two arms as two identical dogs setting out around a single block in opposite directions. They agree in advance that the first dog to arrive at the opposite corner chooses the path for the rest of the walk. If the two dogs walk at the same speed, and the two paths are the same length, they will always arrive at the opposite corner at the same time. If one path is slightly longer than the other, the dog taking that path will always be late arriving, and they will always follow the same route afterward.

"Wait, what's quantum about this? You're just talking about waves and dogs."

"You can describe the basic operation of the interferometer classically, but it also works perfectly well if you send in one photon at a time. That's a system you can only explain with quantum physics."

"Don't you need lots of photons to see the pattern?"

"Sure, but you can repeat the experiment many times, and build up the pattern. You set it up so the two paths have equal length, and repeat it 1,000 times, and you'll see 1,000 photons at one detector. Then you move one mirror a little bit, and repeat the experiment another 1,000 times, and see 700 photons, and so on. If you keep doing this over and over again, you'll trace out the same pattern you see with a bright beam."

"Like the way I measured the wavefunction of a bunny in the backyard by marking its position every night when I went out to chase it?"

"Technically, you measured a probability distribution, the square of a wavefunction, not the wavefunction itself, but yes, that's the basic idea."

"The bunny is most likely to be underneath the bird feeder, you know."

"Yes, because it eats the spilled seed. Try to focus, please."

"Okay."

When we move to a quantum description of the interferometer, talking about it in terms of single photons and wavefunctions, we say that the photon wavefunction splits into two parts at the first beam splitter. In the popular view of many-worlds, you might be tempted to say that this is also when the universe splits in two, as that is when we first acquire a second branch of the wavefunction.

This is tempting, but wrong—the wavefunction has two branches, but they are not "separate universes" at this point. We know this because when we bring them back together, at the second beam splitter, we see interference. The probability of a photon reaching our detector changes as we change the length of one of the paths in the interferometer, indicating that the two separate paths influence each other. The photon goes both ways at the same time, and interferes with itself when we recombine the branches.

We see this interference pattern because the two parts of the wavefunction have a property called **coherence**. "Coherence" is a slippery word, but when we say that two wavefunctions are "coherent," we mean that they behave as if they came from a single source.* In the case of the interferometer, they did come from a single source, and each of the pieces experiences essen-

*This is a very rough definition, and doesn't capture everything—waves from opposite ends of a single large source may not be coherent, for example—but it does get the basic idea across.

tially the same interactions as it goes through the interferometer, so they remain coherent all the way through. The only factor determining whether the interference is in phase or out of phase is the length difference between the two paths.

To turn the two branches of the wavefunction into two separate universes, we need to destroy that coherence. Without coherence, the two pieces of the wavefunction will not interfere to give a detectable pattern, and we will not be able to see any influence of one path on the other. The photon will look like a classical particle in two different universes, passing straight through the beam splitter in one universe, and reflecting off it in the other.

The coherence between the two branches of the wavefunction is destroyed by interaction with a larger environment. To see how this works, let's imagine making the interferometer very long, so that the beams have to pass through a lot of air between the two beam splitters:

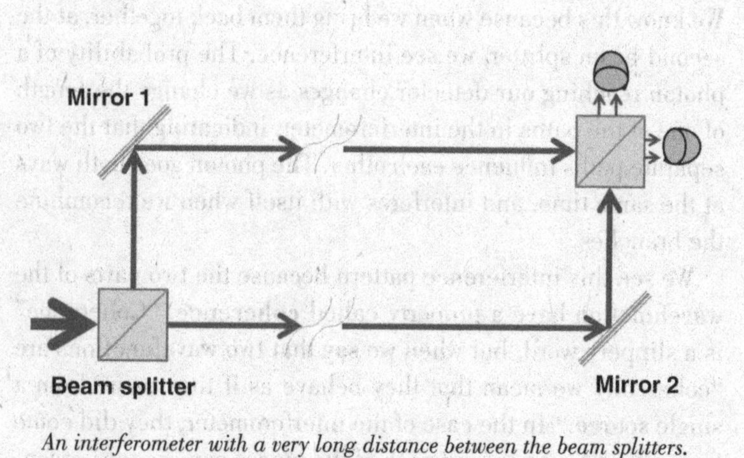

An interferometer with a very long distance between the beam splitters.

In our dog-walking example, we can imagine this longer interferometer as sending our two identical dogs around Central Park in opposite directions. The longer path allows more time for distractions to build up—squirrels to chase, dropped food to eat, horse droppings to roll in. Now the dogs no longer move at the same speed—sometimes they speed up, and sometimes they slow down, depending on what they see. The order in which the two dogs arrive depends not only on the length of the path they follow, but on what they encountered along the way.

The same thing happens with light. As light travels through the interferometer, it occasionally interacts with the atoms and molecules making up the air. Light passing along one arm might encounter a region with more molecules than average, and slow down a little, or it might encounter a region with very few molecules, and speed up a bit. These effects are very small, but they add up over the length of the path, just like the distractions encountered by our walking dogs. The interference pattern is determined not only by the length of the two paths, but also by interactions with the environment in the form of the air that the light passes through.

Interactions with the environment, like distractions around Central Park, are essentially random, and fluctuating—that is, they're different from one place to another, and one time to another. A dog won't find dropped food on the same block every day, and a photon won't interact with a molecule in the same part of the interferometer every time. Interactions shift the pattern, and the shift is different each time we run the experiment. The end result—whether we see light at Detector 2, or which dog wins the race—will be completely random, determined by factors outside our control.

Thanks to these random interactions, the light waves along the two paths are no longer coherent, so we no longer see a clean interference pattern when we bring them back together. Instead we see a pattern that's constantly changing, shifting position millions of times a second. The bright and dark spots blur together, wiping out the pattern. This same effect carries over when we do the experiment with single photons—the two pieces of the wavefunction are no longer coherent. When the photons undergo random, fluctuating interactions with a large environment, the interference pattern is destroyed.

This process of random interactions destroying quantum effects is known as **decoherence**, because it's destroying the coherence between the different parts of the photon wavefunction. Decoherence prevents the various branches of the wavefunction from affecting one another in any detectable way—while we would expect the different branches to come back together and produce interference patterns, we don't see that. Thanks to decoherence, the interference between different branches is random, and thus can't be detected.

"So, now we have two different photons that don't interfere?"

"No, it's more subtle than that. We still have only one photon, but it appears in two different branches of the wavefunction. We don't see any interference because of decoherence."

"It's one photon that doesn't interfere with itself?"

"It interferes with itself, but the interaction with the environment leads to random shifts that make the interference pattern different every time. We can't build up a pattern through repeated measurements, because it's always moving around. The pattern shifts millions of times a second, and you're as likely to get a bright spot as a dark spot, smearing the whole thing out into—"

"An in-between spot?"

"Exactly."

"So, it's like if I tried to measure the bunny wavefunction by marking its position when I went out to chase it, but sometimes I was chasing squirrels instead of bunnies?"

"Well, the squirrels and bunnies both tend to be found under the bird feeder, so that wouldn't make much difference. It's more like trying to measure the bunny wavefunction by recording the position every night, while I keep moving the bird feeder to different spots in the yard."

"Oh. That would be mean. Don't do that."

"I'm not going to. It's just an analogy for the way that decoherence wipes out the interference pattern. Even though each individual photon interferes with itself, the two parts of the wavefunction aren't coherent, so we can't build up a pattern from repeated measurements."

"But if we don't see a pattern, how do we know that there's interference? What's the difference between quantum particles that don't produce an interference pattern and ordinary classical particles?"

"That's exactly the point: there is no difference. Because of the random interactions, the quantum effects are smeared out, and we're left with something that doesn't look like an interference pattern, even though there is interference happening all the time. You end up detecting a photon 50% of the time, exactly as you would if it were a classical particle with no wave properties."

"I don't know. It's kind of Zen, isn't it? 'What is the pattern of one photon interfering?'"

"Hey, that's pretty good."

"Thanks. I have Buddha nature, you know."

THE INFLUENCE OF THE ENVIRONMENT: DECOHERENCE AND MEASUREMENT

You will sometimes see explanations of decoherence describing it as a measurement process, saying things like "when a photon interacts with an air molecule, that's the same as measuring the position of the photon, which destroys the interference pattern." This is almost exactly backward—decoherence isn't a result of measuring the photon through interactions, it's the result of *not* measuring the interaction between the photon and its environment.

This is a subtle point, but it's critical to the modern understanding of decoherence. As we send individual photons through the interferometer, each one interferes with itself, according to a wavefunction that shows some interference pattern. If we could send in thousands of photons, and get the same interaction each time, we could repeat the measurement over and over, and trace out an interference pattern like the first dashed curve shown in the figure on the next page.

We can't guarantee exactly the same interaction between the photon and the environment every time, though, any more than we can guarantee that a Central Park tourist will drop food in the same place every time our dogs go for a walk. As a result, the second photon sent in will interfere with itself according to a different wavefunction. If we could repeat *this* experiment thousands of times, we would trace out a pattern like the second dashed curve in the figure. The third photon has yet another different wavefunction, which would trace out a pattern like the third curve, and so on.

In the end, we get one photon drawn from each of these patterns, and thousands of others, each with peaks in different positions. The cumulative effect is to trace out a pattern that is the sum of many different interference patterns. This is shown as the solid line in the figure, which barely shows any interference at all.

What does this have to do with measuring the environment? Well, when each of the photons we send in interacts with the air, it makes a small change in the state of the environment—an air molecule is moving a little bit faster, or a little bit slower, or the internal state of that molecule is changed in some way. Exactly what happens to the environment depends on exactly what sort of interaction went on, and that, in turn, determines what happens to the photon.

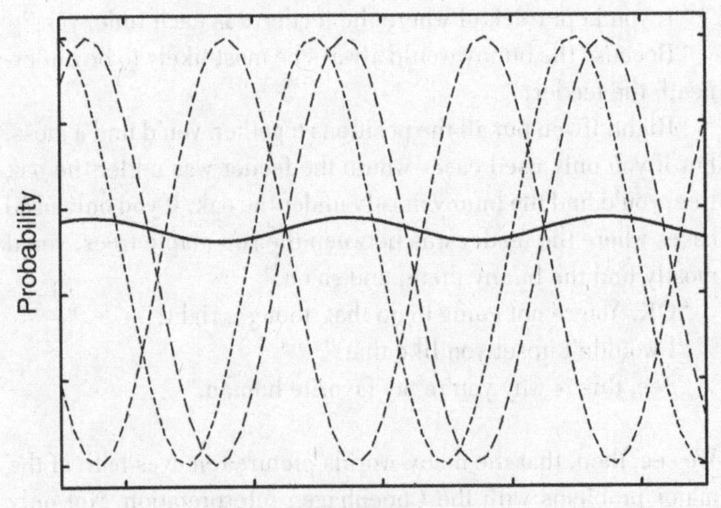

The dashed lines show the interference patterns for photons with three different phases. The solid line represents the sum of several such patterns, showing that the interference pattern is almost completely wiped out.

If we could keep track of everything that happened to the environment—the exact state of every air molecule in the two paths through the interferometer—we could use that information to work out what happened to the photon, and choose to look only at photons whose wavefunctions produce identical patterns. Only a tiny number of photons will have exactly the same result, but if you repeat the experiment often enough, you'll find some—the 159th photon through might produce exactly the same pattern as the first, and then the 1022nd, and the 5,674th, and so on. If you look only at those photons, you'll see them trace out an interference pattern, just as if there were no decoherence.

"So, even if you *were* moving the feeder every night, I could still find the bunny wavefunction?"

"If you kept track of where the feeder was each time, yes."

"Because the bunny would always be most likely to be underneath the feeder?"

"Right. If you put all the positions together, you'd find a mess, but if you only used cases where the feeder was under the oak tree, you'd find the bunny mostly under the oak. If you only used cases where the feeder was between the two maple trees, you'd mostly find the bunny there, and so on."

"OK. You're not going to do that, though, right?"

"I wouldn't upset you like that."

"See, this is why you're my favorite human."

We see, then, that the many-worlds picture removes both of the major problems with the Copenhagen interpretation. Not only does it eliminate the mysterious "collapse" of the wavefunction, it gets rid of the arbitrary division between microscopic and macroscopic in the Copenhagen picture. According to the many-worlds interpretation, macroscopic objects like cats in boxes *do* obey quantum rules, and show superposition and interference effects. We don't *see* those effects because of decoherence caused by interactions with the environment. If we could keep track of every interaction between the cat and its environment, though, we could reconstruct the wavefunction, and verify that quantum mechanics works on every scale.

Of course, it's impossible to keep track of the exact state of every particle making up even the simplest of scientific experiments, let alone all the particles making up a macroscopic object like a piece of steak or a hungry dog. As a result, the process of decoherence happens all the time, and happens extremely quickly. You see decoherence any time you have an object interacting with a larger environment, and all objects are always interacting with their environments. The more atoms you have, the more chances you have for the system to interact with the environment, and the faster decoherence will take place. A piece of steak contains something like 10^{23} atoms, so decoherence will take place extremely quickly, so quickly that we'll never get a chance to see any quantum effects.*

Real objects are extremely difficult to isolate from the environment well enough to be able to see quantum interference, but it can be done. Experimentalists have been able to demonstrate quantum effects using small numbers of particles inside ultrahigh vacuum chambers, often involving components cooled to temperatures close to absolute zero. The largest such experiment involved about a billion electrons in a loop of superconductor, at temperatures within a few degrees of absolute zero. The electrons were placed into a superposition of states corresponding to clockwise and counterclockwise flow around the loop (like a dog walking clockwise and counterclockwise around Central Park at the same time). A billion electrons sounds like a lot, but it's still pretty tiny compared to everyday objects. It serves to demonstrate, though, that with sufficiently careful control of the environment, we can see quantum behavior with large numbers of particles.

GETTING TO REALITY: DECOHERENCE AND INTERPRETATIONS

The idea of decoherence is by no means exclusive to the many-worlds interpretation of quantum mechanics. Decoherence is a real physical process that happens in all interpretations. In the Copenhagen interpretation, it serves as the first step in the process of measurement, selecting the states you can possibly end up in. Decoherence turns a coherent superposition of two or

*This is on top of the fact that the wavelength of a 10-gram piece of steak would be something like 10^{-29} meters, as discussed in chapter 1, making it nearly impossible to measure a steaky interference pattern even if decoherence weren't an issue.

more states (both A and B) into an incoherent mixture of definite states (either A or B). Then some other, unknown, mechanism causes the wavefunction to collapse into one of those states, giving the measurement result. In the many-worlds view, decoherence is what prevents the different branches of the wavefunction from interacting with one another, while each branch contains an observer who only perceives that one branch.

In either case, decoherence is an essential step in getting from quantum superpositions to classical reality. All of the concrete predictions of quantum theory are absolutely identical, regardless of interpretation. Whichever interpretation you favor, you use the same equations to find the wavefunction, and the wavefunction gives you the probabilities of the different possible outcomes of any measurement. No known experiment will distinguish between the Copenhagen interpretation and the many-worlds interpretation,* so which you use is essentially a matter of personal taste. They are just two different ways of thinking about what happens as you move from the probabilities predicted by the wavefunction to the result of an actual measurement.

"So how do I pick my universe?"

"Pardon?"

"I want to be in the universe where I eat steak. What determines which universe I end up in, and how do I change it to give me the one that I want?"

"Well, there's nothing you can do to affect the outcome of a quantum measurement, that we know of. It's completely random,

*Or any of the other interpretations, for that matter. Interpretations of quantum mechanics are sort of "metatheories," each giving a different gloss on the results of an experiment, but not changing the results. Every now and then, you will run across somebody claiming to have experimentally "proved" some interpretation or another, but they're inevitably confused.

whether you want to think of it as a collapsing wavefunction, or just perceiving a single branch of the entire wavefunction of the universe. Either way, you're stuck with random outcomes."

"But I thought you said that in many-worlds the wavefunction always followed the Schrödinger equation? Can't you use that to predict which branch is the real one?"

"You can use it to predict the probabilities of the different branches, but each branch has its own version of you, perceiving its own set of measurement outcomes. Each of them think they're the 'real' branch, and wonder why they didn't end up in a different branch. The theory doesn't say anything about one of them being 'real.'"

"So . . . it's a punt, basically?"

"Sorry, but yes. It's mathematically more elegant, but not really much more satisfying than the Copenhagen interpretation, when you get right down to it. It just sort of pushes the question back a level."

"Interpretations are stupid."

"You're not alone in thinking so, but we're stuck with them for the moment."

"Well, I don't like them. I want to eat steak. If quantum measurement won't help me eat steak, then I don't want any more to do with it."

"Interpretations aren't the whole of quantum theory, by any stretch. There are only a few people who spend their time working on that stuff. Most physicists don't bother worrying about interpretations, and use quantum mechanics to do useful stuff instead."

"If measurement is random, how can you do anything useful with it?"

"Well, that's the next chapter."

CHAPTER 5

Are We There Yet? The Quantum Zeno Effect

I've had a long, annoying day at work, running from one committee meeting to another, and I come home with a pounding headache. I have an hour or so until Kate gets home, and all I want is a nap. Emmy is ecstatic to see me, and does the Happy Dance all over the living room.

"Hooray! You're home! Yippeee!!!" She's wagging her tail so hard she almost loses her balance. This happens every afternoon when I come home.

"It's good to see you, too."

"Let's do something fun! Let's play fetch! Let's go for a walk! Let's play fetch on a walk!"

"Let's let me take a nap." She stops bounding immediately, and looks crestfallen. Her ears and tail droop.

"No walk?"

"Not right now," I say, lying down on the couch. "Let me sleep for half an hour, and then we'll do something fun."

"Promise?"

"I promise. Now be quiet. The sooner I get to sleep, the sooner we'll do something fun."

"Oh. Okay."

I lie down and get comfortable on the couch. I'm just starting to settle in for my nap, when a cold, wet nose pokes me in the face.

"Are you asleep?"

"No, I'm not asleep."

"Oh." A minute passes.

Poke. "Are you asleep?"

"No." Another minute passes.

Poke. "Are you asleep?"

"No!" I sit back up. "And I'm never going to get to sleep if you don't stop poking me with your nose and asking that question."

"Why not?"

"Every time you poke me, you wake me back up, and I have to start over again. If you keep waking me up, I don't get all the way asleep, and you don't get to do anything fun."

"Oh." She brightens up. "Hey, it's just like the Zero Effect!"

"The what?"

"You know. The paradox with the guy who can't catch the turtle because he has to go half the way there, and then another half, and so on, so he never gets anywhere."

"You mean Zeno's paradox. Zeno, with an *n* as in 'nap.' The *Zero Effect* is a movie with Bill Pullman and Ben Stiller."

"Whatever. I don't spell so good."

"Anyway, what you're thinking of is the ***quantum* Zeno effect**, and yes, this is kind of like that. If you have a system that's moving from one state to another, with the probability of being in the second state increasing over time, you can prevent the state change by repeated measurements. Every time you measure it to be in the first state, you restart the process."

"Right. So when I ask if you're asleep, I collapse your wavefunction back to the 'awake' state, and you need to start napping again."

"Or you find yourself perceiving the branch of the wavefunction in which I'm awake again, in the many-worlds picture. But yes, that's the basic idea, and a good analogy."

"I'm a philosophical dog!"

"Yes, you're very smart. Now shut up and let me sleep."

"Okay. I won't ask if you're asleep anymore."

"Thank you."

I settle back down onto the couch, and start to feel warm and cozy, and feel myself drifting off . . .

Poke. "Are you awake?"

Whether you prefer the Copenhagen interpretation, many-worlds, or one of the many others, *something* happens when you make a measurement. Whether you think this involves the physical collapse of a wavefunction, or just limiting your perception to a single branch of an expanding and evolving wavefunction, measurement is an active process. Before you measure an object's state, it exists in a quantum superposition of all possible states, while immediately after the measurement, you observe one and only one state.

In this chapter, we'll look at the most dramatic consequence of active measurement, the "quantum Zeno effect." We'll see that making repeated measurements of a quantum particle can prevent it from changing its state. We can also use the quantum Zeno effect to detect the presence of objects without hitting them with even a single photon of light.

YOU CAN'T GET ANYWHERE FROM HERE: ZENO'S PARADOX

The name of the effect is a reference to the famous paradoxes of the Greek philosopher Zeno of Elea, who lived in the fifth century B.C.E. There are several different versions of the paradoxes, but all of them purport to show that motion is impossible.

Here's a modern canine version of the argument: in order to reach a treat on the far side of the room, a dog first needs to cross half the width of the room, which takes a finite time. Then, she needs to cross half of the remaining distance, which takes a finite time, and then half of the remaining distance, and so on.

The distance across the room is divided into an infinite number of half steps, each requiring a finite time to cross. If you add together an infinite number of steps, each taking a finite time, it should take an infinite amount of time to cross the room. Thus, it's impossible for the poor dog to ever get all the way to the tasty treat.

Happily for hungry dogs everywhere, there's a mathematical solution to the apparent paradox: as the distance gets smaller, the time required to cross it also gets smaller. If it takes one second to cross half the width of the room, it takes half a second to cross the next quarter, and a quarter of a second to cross the next eighth, and so on. Adding together all those times, we find that:

$$1 \text{ s} + 1/2 \text{ s} + 1/4 \text{ s} + 1/8 \text{ s} + \ldots = 2 \text{ s}$$

The total time is the sum of an infinite number of terms, but the terms get smaller as you go. Mathematicians learned how to add this sort of series when calculus was invented in the seventeenth and eighteenth centuries. The infinite sum gives a finite result: the dog crosses the room in two seconds. Motion is possible after all, and a good dog can always reach her treats.*

WATCHED POTS AND MEASURED ATOMS: THE QUANTUM ZENO EFFECT

The quantum Zeno effect uses the active nature of quantum measurement to prevent a quantum object (like an atom) from

*While the summing of infinite series is accepted as the resolution of Zeno's paradox by physicists and engineers and most mathematicians, some philosophers do not accept this as a sufficient resolution of Zeno's paradox (*Stanford Encyclopedia of Philosophy*). This just proves that philosophers are crazier than mathematicians, or even cats.

moving from one state to another, by making repeated measurements. If we measure the atom a very short time after the transition starts, it will most likely be found in the initial state. The act of measuring the atom projects it back into the initial state, as we saw in chapter 3, and the transition starts over.

If we keep measuring the state of the atom, we keep putting it back where it started. The atom is in a predicament reminiscent of Zeno's paradox—taking an infinite number of steps toward some goal, but never getting there.* As the old saying has it, a watched pot never boils, at least as long as it's a quantum pot.

This is dramatically different from classical physics. Measuring the state of a classical object does not change the state—if a pot of water is 50% of the way to boiling when the measurement is made, it's still 50% of the way to boiling after the measurement. The quantum Zeno effect works only because of the active nature of quantum measurement—the water in a quantum pot is either boiling or not boiling. If you find that it isn't boiling, you need to start over, as if you had never heated it.

The definitive quantum Zeno effect experiment was done in 1990 by Wayne Itano in Dave Wineland's group at NIST in Colorado, using beryllium ions. Ions are just atoms with one electron removed, and like all atoms, they have a collection of allowed energy states, which they move between by absorbing or emitting light. Itano's experiment collected a few thousand beryllium ions, and made them move slowly from one state to another by exposing them to microwaves.

Left unmeasured, the ions took 256 milliseconds to complete the transition from State 1 to State 2.† Their state during this

*A better Greek literary allusion might be the myth of Sisyphus, who was condemned to spend eternity pushing a boulder up a hill, only to have it slip free and roll back to the bottom again. The name "Sisyphus effect" was used for something else, though, so this is called the quantum Zeno effect.

†A quarter of a second seems pretty fast to humans or dogs, but it's really slow for an atom. Atoms usually change states in a few billionths of a second.

process was described by a wavefunction with two parts, corresponding to the probability of finding the atom in State 1 and State 2. At the start of the experiment, the atoms were 100% in State 1, and at the end, they were 100% in State 2. In between, the probability of State 2 steadily increased, while the probability of State 1 steadily decreased.

The experimenters measured the state of the ions using an ultraviolet laser with its frequency chosen so that an ion in State 1 would happily absorb light, while ions in State 2 would not absorb any light. Ions in State 1 absorbed photons from the laser and re-emitted them a few nanoseconds later, making a bright spot on a camera pointed at the ions. Ions in State 2, on the other hand, produced no light when illuminated by the laser. The total amount of light reaching the camera, then, was a direct measurement of the number of ions in State 1.

To demonstrate the quantum Zeno effect, the NIST group trapped a large number of ions, all in State 1. Then they turned on the microwaves, waited 256 milliseconds, and pulsed on the laser. None of the ions produced any light, indicating that 100% of the sample had moved to State 2, as expected. Then they repeated the experiment, with two laser pulses: one after 128 ms (halfway through the move to State 2), and one after 256 ms. In this case, they saw half as much light after 256 ms, indicating that only 50% of the sample had made the transition to State 2.

The decreased probability is explained by the quantum Zeno effect. The laser pulse halfway through measured the state of the ions. Many of them were found in State 1, and the measurement destroyed the State 2 part of the wavefunction. These atoms were now 100% in State 1, so the transition had to start over, with the probability of State 2 increasing slowly. After another 128 ms, the probability of finding the ions in State 2 was only 50%.

The probability of moving from State 1 to State 2 decreased further with more measurements. With four pulses (at 64, 128, 192, and 256 ms), only 35% of the atoms made the transition.

With eight pulses, only 19% made the transition. With a total of 64 laser pulses over the full experimental interval (one every 4 ms), fewer than 1% of the atoms made the transition. All of these probabilities were in excellent agreement with the theoretical predictions of the quantum Zeno effect, as shown in the figure below.

"So, when you make a measurement, the ion absorbs a photon, and that collapses the wavefunction?"

"Actually, the ion doesn't need to absorb a photon at all. The Wineland group repeated the experiment starting with the ion in State 2. In that case, the ion starts out in the 'dark' state, and doesn't absorb any photons during the measurements. They still got the same result—the probability of making the transition from State 2 to State 1 decreased with more measurements, exactly as predicted."

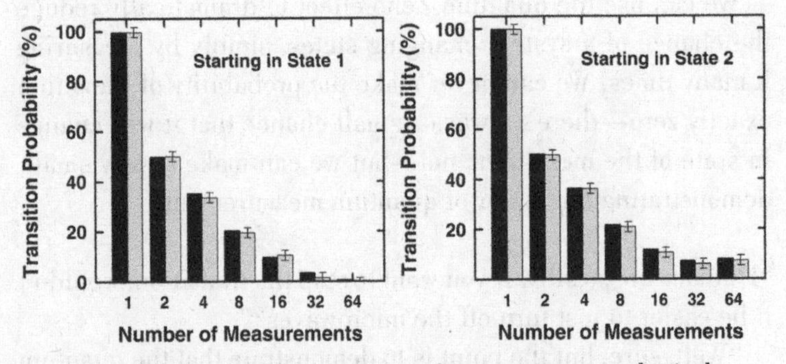

The probability of making a transition from one state to another in the quantum Zeno effect experiment done by the Wineland group (W. M. Itano, D. J. Heinzen, J. J. Bollinger, and D. J. Wineland, Phys. Rev. *A 41, 2295–2300 [1990], modified and reprinted with permission). Black bars are the theoretical prediction, gray bars are the experimental result, with error bars showing the experimental uncertainty. The probability of changing states decreases as the number of measurements increases, whether the ions start in State 1 or State 2.*

"Wait, not absorbing a photon is the same as absorbing a photon?"

"When it comes to thinking of the photons as measurement tools, yes. It's just like the treat in two boxes—if you open one of the boxes, and find it empty, you know the treat has to be in the other box. That determines the state of the treat just as if you opened the box and found a treat there."

"It's not as much fun, though, because I don't get the treat."

"Yes, well, your life is very difficult."

The quantum Zeno effect does not depend on a particular interpretation of quantum mechanics. It's easier to discuss what's going on using the Copenhagen language of wavefunction collapse, but we can equally well describe it in terms of the many-worlds interpretation. In the many-worlds picture, new branches of the wavefunction appear at each measurement step, but we are more likely to perceive the higher probability branch. The probability of seeing a state change is the same in both interpretations.

We can use the quantum Zeno effect to dramatically reduce the chance of a system changing states, simply by measuring it many times. We can never make the probability of transition exactly zero—there's always a small chance that it will change in spite of the measurements—but we can make it very small, demonstrating the power of quantum measurement.

"Humans are so silly. If you want to stop the transition, wouldn't it be easier to just turn off the microwaves?"

"Well, sure, but the point is to demonstrate that the quantum Zeno effect is real. It's not interesting because it can stop ions from changing states; it's interesting because of what it tells us about quantum physics."

"Yeah, but what good is it? Can it do anything useful?"

"Well, you can use it to detect objects without having them absorb any light."

"Objects . . . like bunnies?"
"Sure, hypothetically."
"Oooh! I like the sound of that!"

MEASURING WITHOUT LOOKING: QUANTUM INTERROGATION

The quantum Zeno effect can be exploited to do some amazing things. A collaboration between the University of Innsbruck and Los Alamos National Laboratory has demonstrated that it's possible to use light to detect the presence of an absorbing object *without having it absorb any photons,* by using the quantum Zeno effect to stop a photon moving from one place to another.

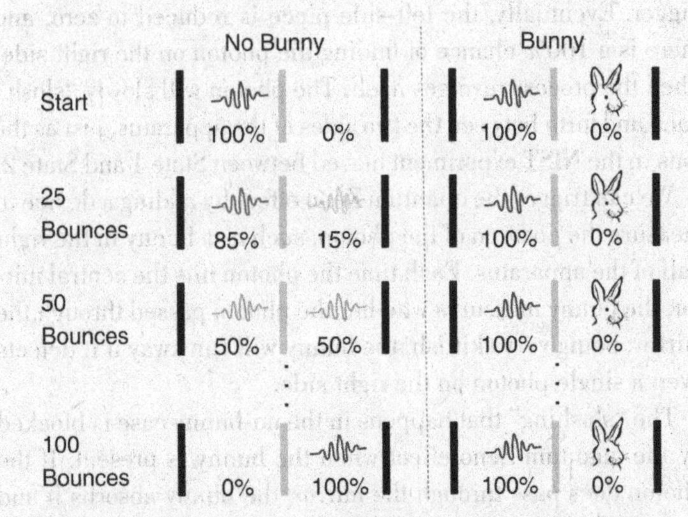

We start with a photon on the left-hand side of the apparatus, bouncing back and forth between two mirrors. There is a small chance of the photon leaking through the central mirror, so over time the photon will shift into the right-hand side of the apparatus. If there is an absorbing object (a bunny, say) on the right-hand side, though, it will prevent the photon from moving, through the quantum Zeno effect.

In the future, this technique may be used to study the properties of quantum systems that are too fragile to survive absorbing even a single photon.

Here's a simplified version of this **quantum interrogation** experiment: imagine that we have a single photon bouncing back and forth between two perfect mirrors. Halfway between those two, we insert a third mirror that's not quite perfect.

The wavefunction for this system has two pieces, one corresponding to finding the photon in the left half of the apparatus, and the other corresponding to finding the photon in the right half. If we start the experiment with a single photon in the left half, we find that over time, it will slowly move into the right half. Each time the photon hits the imperfect central mirror, there's a small chance that it goes through, so the left-side piece of the wavefunction gets a little smaller, and the right-side piece gets bigger. Eventually, the left-side piece is reduced to zero, and there is a 100% chance of finding the photon on the right side. Then the process reverses itself. The photon will slowly "slosh" back and forth between the two sides of the apparatus, just as the ions in the NIST experiment moved between State 1 and State 2.

We can trigger the quantum Zeno effect by adding a device to measure the position of the photon, such as a bunny in the right half of the apparatus. Each time the photon hits the central mirror, the bunny measures whether the photon passed through the mirror: being very skittish, the bunny will run away if it detects even a single photon on the right side.

The "sloshing" that happens in the no-bunny case is blocked by the quantum Zeno effect when the bunny is present. If the photon does pass through the mirror, the bunny absorbs it and flees. The photon no longer exists, so its wavefunction is zero, and nothing changes after that. If it doesn't make it through, the photon is definitely on the left-hand side, and the wavefunction is put back in the initial photon-on-the-left state, and everything starts over.

The quantum Zeno effect lets us do what any dog wants to: determine whether there's a bunny in the apparatus without scaring it off. We start with a photon on the left side, wait long enough for it to move over to the right side, and then look at the left side of the apparatus. If there's no photon there, there's no bunny on the right, either because the bunny absorbed the photon and ran off, or because there never was a bunny and the photon has "sloshed" over to the right. If the photon is still in the left-hand side of the apparatus, we know that not only was there a bunny, but it is still there, and has not absorbed even a single photon of light.

There is always a chance that the photon will make it through and scare the bunny away, but we can make this chance as low as we like, by decreasing the probability that the photon will leak through the mirror. We'll have to wait longer to complete the measurement, as the time required for the photon to "slosh" into the right side will increase, but the chances of successfully detecting the bunny improve dramatically. If the photon needs to bounce back and forth on the left-hand side 100 times before it "sloshes" to the right, the probability of detecting a bunny *without scaring it off* is 98.8%. If you repeated the experiment 1,000 times, only 12 bunnies would be scared off.

"Oooh! So, all I need to do is get some big mirrors . . ."

"No. You are not setting this experiment up in the backyard."

"But I can use the quantum Zeno effect to sneak up on the bunnies . . ."

"No. Just . . . No. You are not putting great big mirrors across the yard, and that's final."

"Awww . . ."

Quantum interrogation hasn't been used to catch bunnies, but it has been demonstrated experimentally using polarized photons, by physicists in Innsbruck, Los Alamos, and Illinois. Quantum interrogation allows you to do some incredible things—taking pictures of objects without ever bouncing light off them, for example. This probably isn't useful for spy purposes (unless you can somehow get your enemies to obligingly store their secrets between two mirrors), but it might be essential for probing fragile quantum systems like large collections of atoms in superposition states that can't survive the absorption of a photon.

Whether you think of it in terms of collapsing wavefunctions, or a single expanding wavefunction undergoing decoherence, the quantum Zeno effect is a dramatic demonstration of the strange nature of quantum measurement. Unlike classical measurement, the act of measuring a quantum system changes the state of that system, leaving it in only one of the allowed states, which is very different than what we expect classically. With a clever arrangement of the experimental situation, this can be exploited to prevent a system from changing states, or even to extract information from a system without interacting with it directly.

"That's really interesting. Weird, but interesting."

"Thanks."

"Now, if you'll excuse me, I need to go look in my bowl."

"Why is that?"

"Well, I'm going to use the Zeno effect to get more food. I figure, if I keep measuring my bowl to be full of kibble, I'll always have kibble, no matter how much I eat. That will be fun."

"Of course, if you keep measuring your bowl to be empty, it'll always be empty, and you'll never have kibble."

"Oh. That would be bad. I didn't think of that."

"Anyway, you'd need to have some natural quantum process that caused kibble to appear in the bowl for that to work. Things aren't going to appear for no reason, just because you want to measure them."

"Well, you sometimes put kibble in my bowl, right? And you're a natural process."

"In a manner of speaking."

"So, how about putting some kibble in my bowl?"

"Oh, all right. It's almost dinnertime. Come on."

"Oooh! Kibble!"

CHAPTER 6

No Digging Required: Quantum Tunneling

We're sitting in the backyard, enjoying a beautiful sunny afternoon. I'm lying on a lounge chair reading a book, and Emmy is sprawled out on the grass, basking in the sun and keeping an eye out for squirrel incursions.

"Can I ask a question?" she asks.

"Hmm? Sure, go ahead."

"What do you know about tunneling?"

"Tunneling, eh?" I put my book down. "Well, it's a process by which a particle can get to the other side of a barrier despite not having enough energy to pass over the barrier."

"Barrier? Like a fence?"

"Well, metaphorically, at least."

"Like the fence between this yard and the next?" She looks really hopeful.

"Oh. Is *that* what this is about?"

"There are bunnies over there!" She wags her tail for a minute, then looks crestfallen. "But I can't get to them."

"True, but I don't think tunneling is the answer. It works for small particles, but wouldn't work for a dog."

"Why not?"

"Well, you can think of a barrier in terms of potential and kinetic energy. For example, right now, all your energy is potential energy, because you're not moving. But you could start moving, say, if you took off after a squirrel, and turned that potential into kinetic energy."

"I'm very fast. I have lots of energy."

"Yes, I know. You're a great trial to us. Anyway, whether you're sitting still, or moving, you have the same total amount of energy. It's just a question of what form it's in."

"Okay, but what does this have to do with the fence?"

"Well, you can think of the fence as being a place where you can only go if you have enough energy. For you to be at the spot where the fence is, you would have to jump very high or else occupy the same space as the fence, and either would take an awful lot of energy."

"I can't jump that high. That's why I can't get the bunnies."

"Right. You don't have enough energy to get over the fence. And because you don't have enough energy, you can't end up in the neighbors' yard, and everybody is much happier that way, believe me."

"Except me." She pouts.

"Yes, well, except you." I scratch behind her ears by way of apology. "Anyway, quantum mechanics predicts that even though you don't have enough energy to go over the fence, there's still a chance that you could end up on the other side. You could just sort of . . . pass through the fence, as if it weren't there."

"Like the bunnies do!"

"Pardon?"

"The bunnies. They go back and forth through the fence all the time."

"Yes, well, that's because they fit between the bars of the fence. It has nothing to do with quantum tunneling." I stop for a moment. "Of course, it's not a bad analogy. The bunnies don't have enough energy to go over the fence, either, but they can go through it, and end up on the other side. Which is sort of like tunneling."

"So how do I tunnel through the fence?"

"Well, you could eat fewer treats, and get skinny enough to pass between the bars like the bunnies do."

"I don't like that plan. I'm a good dog. I deserve the treats I get."

"And you get the treats you deserve. The other option would be quantum tunneling through the fence, but quantum tunneling isn't something you *do,* it's something that just happens. If you send a whole bunch of particles at the barrier, a small number of them will show up on the other side. But *which* ones go through is completely random. It's all about probability."

"So, I just need to run at the fence enough times, and I'll end up on the other side?"

"I wouldn't try it. The probability of a particle tunneling through a barrier depends on the thickness of the barrier and the quantum wavelength of the particle. The probability of a fifty-pound dog passing through a half-inch aluminum barrier would be something like one over *e* to the power of ten to the thirty-six. Do you know what that is?"

"What?"

"Zero. Or near enough to make no difference. So don't go throwing yourself at the fence."

She's quiet for a minute.

"Anyway, I hope that answers your question." I pick my book back up.

"Sort of."

"Sort of?"

"Well, the quantum stuff was interesting, and all, but I was thinking of classical tunneling."

"Classical tunneling?"

"I was going to dig a hole under the fence."

"Oh."

"It's a good plan!" She wags her tail enthusiastically, and looks very pleased with herself.

"No, it's not. Only bad dogs dig holes."

"Oh." Her tail stops, and her head droops. "But I'm a good dog, right?"

"Yes, you're a very good dog. You're the best."

"Rub my belly?" She flips over on her back, and looks hopeful.

"Oh, okay . . ." I put my book back down, and lean over to rub her belly.

"Tunneling" is one of the most unexpected quantum phenomena, where a particle headed at some sort of obstacle—say, a dog running toward a fence—will pass right through it as if it weren't there. This odd behavior is a direct consequence of the underlying wave nature of quantum particles seen in chapter 2.

In this chapter, we'll talk about the essential physics concept of energy, and how energy determines where particles can be found. We'll see that the wave nature of matter allows quantum particles to turn up in places that classical physics says they can't reach, passing into or even through solid objects. We'll also see how tunneling lets scientists build microscopes that can study the structure of matter, making possible revolutionary developments in biochemistry and nanotechnology.

THE ABILITY TO GET THINGS DONE: ENERGY

In order to explain quantum tunneling, we need to first talk about the classical physics of **energy**. While the term "energy" has passed from physics into more general use, its physics meaning is slightly different from its everyday, conversational use.

A one-sentence definition of the term "energy" in physics might be: "The energy content of an object is a measure of its ability to change its own motion or the motion of another object." An object can have energy because it is moving, or because it is held stationary in a place where it might start moving. Every object has some energy simply because it has mass (Einstein's $E = mc^2$) and because its temperature is above absolute zero.* All of these forms of energy can be used to set a stationary object into motion, or to stop or deflect an object that is moving.

The most obvious form of energy is **kinetic energy,** the energy associated with a moving object. The kinetic energy of an object moving at an everyday sort of speed is equal to half its mass times the velocity squared, or as it's usually written:

$$KE = \frac{1}{2} mv^2$$

Kinetic energy is always a positive number, and increases as you increase either the mass or the speed. A Great Dane has more kinetic energy than a little Chihuahua moving at the same speed, while a hyperactive Siberian husky has more kinetic energy than a sleepy old bloodhound of the same mass. Kinetic energy is similar to momentum, but it increases faster as you increase the velocity, and unlike momentum, it doesn't depend on the direction of motion.

Objects that are not already moving have the potential to start moving due to interactions with other objects. We describe this as **potential energy.** A heavy object on a table has potential energy: it's not moving, but it can acquire kinetic energy if a hyperactive dog bumps into the table and it falls on her. Two magnets held close to each other have potential energy: when released, they'll either rush together or fly apart. A dog always has potential energy, even when sleeping: at the slightest sound, she can leap up and start barking at nothing.

*Temperature measures the energy due to the motion of the individual atoms making up an object, and "absolute zero" is the imaginary temperature at which that motion would cease. No real object can be cooled all the way to absolute zero, though, and even if one could it would still have zero-point energy, as discussed in chapter 2 (page 49).

Energy is essential to physics because it's a "conserved quantity": the law of **conservation of energy** says that while energy may be converted from one form to another, the total amount of energy in a given system does not change. This turns some difficult problems into bookkeeping exercises: the total energy (kinetic plus potential) has to be the same at the end of the problem as at the beginning, so whatever energy is left over when you subtract the final potential from the total has to be kinetic energy.*

To get a better feel for how energy works, let's think about a concrete example: a ball thrown up in the air. As any dog knows, what goes up must come down, and a ball that's thrown up with some initial velocity will slow down, stop, and then fall back down. You can see this in the figure on the next page, which shows the height of a ball at a series of regularly spaced instants. At low heights, the ball is moving fast, and covers a lot of ground from one picture to the next. Near the top of its flight, the ball moves very little, and at the very peak, it's perfectly still for a split second.

We can describe this flight in terms of energy. An instant after the toss, the ball is moving, so it has lots of kinetic energy, but it's near the ground, and has no potential energy. The total energy is thus equal to the kinetic energy. We can think of this as being a kind of energy supply, like a jar full of treats, shown by the black bar in the figure. As the ball moves upward, its kinetic

*Potential energy is generally much easier to calculate than kinetic energy. Potential energy usually depends only on the positions of the interacting objects, while the kinetic energy depends on the velocity, which depends on what has happened in the recent past. The easiest way to tackle an energy problem is usually to calculate the potential energy using the position, and find the kinetic energy by process of elimination. For example, when a roller coaster pauses at the top of a big hill, we know that all of its energy is potential energy. Later on, we can easily calculate the potential energy from the height of the track, and that lets us find the kinetic energy (and thus the speed) without needing to know what happened in between.

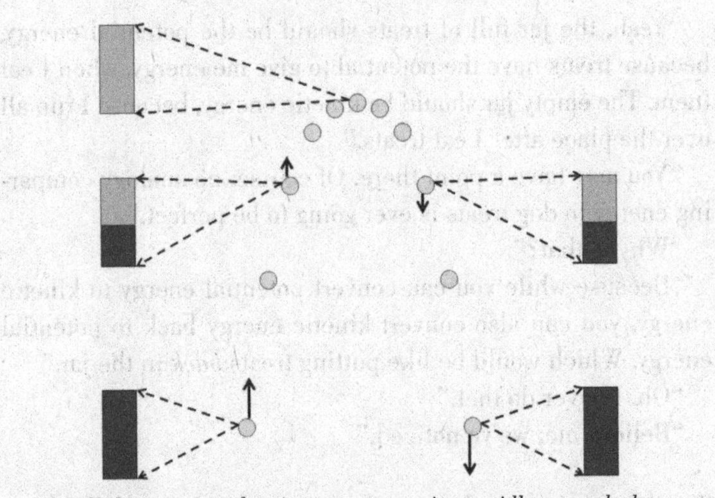

A ball thrown up in the air starts out moving rapidly upward, slows due to gravity, and turns around and falls back. The pictures show the position of the ball at regular intervals. The bars show the energy of the ball, with black indicating kinetic energy and gray indicating potential. Near ground level, all of the energy is kinetic energy, while at the peak of its flight, all of the energy is potential energy.

energy decreases (because it's not moving as fast), and its potential energy increases (because it's higher off the ground). The kinetic energy level drops, replaced by potential energy (shown in gray), but the total energy remains the same.

At the peak of its flight, the ball has potential energy, but no kinetic energy, because for a split second, it's not moving at all. On the way back down, it goes through the same process in reverse: it starts with potential energy but no kinetic energy, and ends up with kinetic energy (the same amount it started with) but no potential energy.

"You've got this backward, you know."

"I do?"

"Yeah, the jar full of treats should be the potential energy, because treats have the potential to give me energy, when I eat them. The empty jar should be kinetic energy, because I run all over the place after I eat treats."

"You may have a point there. Of course, no analogy comparing energy to dog treats is ever going to be perfect."

"Why is that?"

"Because while you can convert potential energy to kinetic energy, you can also convert kinetic energy back to potential energy. Which would be like putting treats *back* in the jar."

"Oh. I never do that."

"Believe me, we've noticed."

We also see from looking at the energy of a ball in flight that energy limits the motion of the ball. The ball starts with some total energy, all kinetic, and as it goes up, it converts that to potential energy. Once the initial kinetic energy has been turned into potential, the ball stops moving. The ball can't go beyond a certain maximum height, because that would require its total energy to increase, and that can't happen.* The maximum height the ball can reach with a given amount of energy is called the "turning point," because the ball reverses direction at that point. Heights above the turning point are "forbidden," because the ball doesn't have enough energy to reach them.

*The total energy of an object *can* be increased, by adding energy from some other source, in the same way that a dog's treat jar can be refilled by a friendly human. The extra energy does not come for free, though—the energy of the outside object has to decrease, in the same way that a human's bank balance will decrease in order to supply the treats. The total energy of the entire universe—balls, dogs, treats, and humans—is a constant, and has not increased or decreased in the fourteen billion years since the Big Bang.

FOLLOW THE BOUNCING WAVEFUNCTION: A QUANTUM BALL

The thrown ball is a simple example of energy in action, and thinking about it in energy terms may not seem that helpful. Energy analysis can be applied to many more complicated systems, though, including situations that can *only* be described mathematically using energy. As a result, energy is one of the most important tools that physicists have for understanding the world.

Energy is especially important in talking about quantum mechanics. As we saw in chapter 2, quantum particles do not have a well-defined position or velocity, so there's no way to keep track of those properties as we do with a classical system. Conservation of energy still applies, though, so we can understand quantum systems by looking at their energy. In fact, the Schrödinger equation uses the potential energy of a quantum object to predict what will happen to the wavefunction of that object, so every calculation done in quantum mechanics is fundamentally about energy.

We can see how energy relates to wavefunctions by imagining a quantum ball thrown in the air. We can predict some features of its wavefunction just by using what we know about its energy. Kinetic energy is similar to momentum, and we know from chapter 1 (page 10) that the momentum determines the wavelength. Near the ground, where the kinetic energy is high, the ball should have high momentum and thus the wavefunction should have a short wavelength. Higher up, where the ball is moving slowly, the ball has low momentum, and the wavefunction should have a longer wavelength. We also expect the probability of finding the ball above the turning point to be zero, because the ball should never go higher than allowed by its initial energy.

We can calculate the wavefunction for this system, and we find a probability distribution that looks like this:

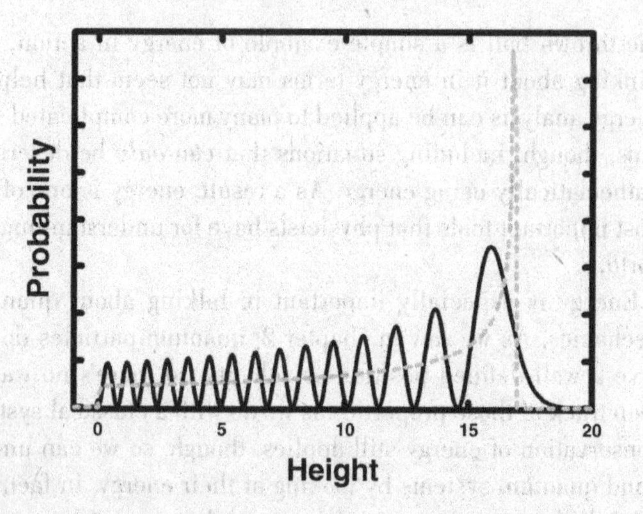

The solid curve shows the probability distribution for finding a quantum particle at a given height. The dashed curve is the probability distribution for a classical particle.

Looking at this graph, we see more or less what we expect. The probability distribution oscillates more rapidly at low elevation (on the left), than at higher elevation. On closer inspection, though, we notice something strange: the probability does not go to zero exactly at the classical turning point (at about 17 units of height, where the dashed curve goes to zero). It drops off to zero, but there's a range of heights above the turning point where the probability is still significant. There's some probability of finding the ball at heights it should never be able to reach!

Why doesn't the probability go to zero right at the turning point? Well, if it did, there would be a sudden change in the wavefunction at that point. We know from chapter 2 (page 47), though, that making a sudden change in the wavefunction requires adding together a huge number of wavefunctions with different wavelengths. Many wavelengths means a large uncertainty in the momentum, and therefore a large uncertainty in kinetic energy. But we don't *have* a large uncertainty in the energy—we know how hard we threw the ball. A small energy uncertainty leaves us with a large uncertainty in the position of the turning point, which means no sharp changes in the wavefunction and a wavefunction that extends into the forbidden region.

The ball can't both have a well-defined energy and turn around exactly where classical physics says it should. If we want a small uncertainty in the energy, we have to accept more uncertainty in the position, and that means that there will be some chance of finding the ball at higher elevation than classical physics allows.

"Why do the wiggles get bigger near the top?"

"I already said that. The ball is moving slower, so the wavelength gets longer."

"Yeah, but they get taller, too."

"Oh, that. That's also because the ball is slowing down. The ball spends more time near the top of its flight where it's moving slowly than it does near the bottom where it's moving fast, so there's a higher probability of finding it near the top. You see the same thing with a classical ball, if you work out the probability distribution—that's the dashed curve."

"So, wait, the most likely position for the ball is way up in the air?"

"Yep. You can see that by looking at the figure on page 125 showing the ball in flight. There are a lot more pictures of the ball at high elevations than at low ones. That's why when we play fetch, you usually catch the ball when it's at the top of its flight—it's not moving very fast, so it's easier to get."

"I'm very good at catching things. Oooh! We should go play fetch! That's fun!"

"After we finish this chapter, okay? I haven't talked about tunneling yet."

"Oh, right. Let's talk about tunneling. Passing through solid objects is even more fun than fetch."

"Don't get your hopes up too much . . ."

LIKE IT'S NOT EVEN THERE: BARRIER PENETRATION AND TUNNELING

How does uncertainty in the turning point lead to particles passing through solid objects? Well, the inside of a solid object is a forbidden region—interactions between the atoms making up the two objects make the potential energy enormous for one object inside another. It's a little like trying to stick a second dog into a kennel that already contains one unfriendly dog—you'll have a hard time getting the second dog in there, and if you do, you'll see a lot of extra energy, in the form of growling and barking and snapping.

In quantum mechanics, though, wavefunctions can extend into forbidden regions, and that works even for solid objects—there's a tiny probability of finding one object inside another. Better yet, if the forbidden region is very narrow, quantum mechanics predicts a small probability of one object passing *through* the other, even though it doesn't have enough energy to make it into the forbidden region, let alone to the other side.

The simplest example of this is an electron hitting a thin piece of metal, where the potential energy is much higher. Classical physics tells us that the kinetic energy of an electron outside the metal determines what happens when it reaches the edge of the metal. If the electron's kinetic energy is large, it can convert most of its energy to potential, and still have kinetic energy left to move through the metal. If the kinetic energy outside is less than the potential inside the metal, though, there's no way the electron can enter without increasing its total energy. The edge of the metal becomes a turning point, and the metal is a forbidden region: an electron coming in from the left bounces off the surface and goes back where it came from. An electron coming in from the right bounces off the other surface in the same way.

According to quantum mechanics, though, we can't have a sharp turning point at the edge of the metal. As with the thrown ball, the electron's wavefunction extends into the forbidden region where the potential energy is greater than the energy of the incoming particles. There's some probability of finding the electrons inside the metal, even though classical physics says it's forbidden. The probability of finding a particle in the forbidden region is highest near the edge, and decreases rapidly as you move farther in. If the forbidden region extends over a long enough distance, the probability drops to zero,* and that's the end of it.

For a very narrow barrier, though, there is some probability of finding the electrons at the opposite edge of the forbidden region from where they entered. Beyond that point, they're no longer forbidden to be there—they're back out in empty space, and move off with the same energy they had at the start. Somebody watching the experiment would see a tiny fraction of the incoming particles—one in a million, say—simply pass through the barrier as if it weren't even there. This is called **tunneling**, because the electrons have passed the forbidden region even though it's impossible for them to be inside it. In a sense,

*Strictly speaking, the probability is never exactly zero—the mathematical function describing the probability is an exponential, and while it gets closer to zero as the electron moves into the barrier, it never gets all the way there. Quantum physics predicts a tiny probability that a ball thrown in the air will tunnel through the forbidden region that starts at its classical maximum height, and end up on the Moon. That's not a good bet, though—the probability is so small that it's indistinguishable from zero, for all practical purposes.

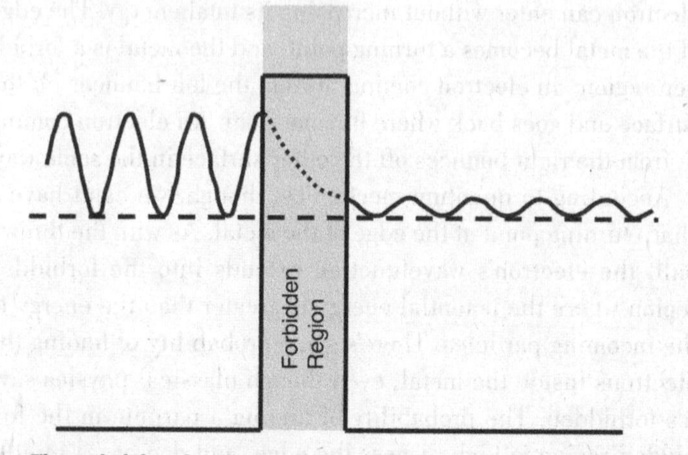

The probability distribution for an electron coming in from the left, hitting a barrier where the potential energy of an electron would increase above the total energy of the electron. The probability drops off rapidly inside the forbidden region, but does not reach zero, so there is some probability of finding the electron to the right of the barrier.

they've ducked under the barrier, like a bad dog who tunnels under a fence.

The wavefunction for this situation is shown above. On the left, we have an incoming electron with some momentum and energy, represented by a wave with a well-defined wavelength.* When the electron reaches the edge of the metal, it enters the forbidden region, and the probability decreases rapidly. It doesn't get to zero before it reaches the right edge of the forbidden region, though, so it emerges there as another wave with the same wavelength as on the left.

The smaller height of the wave to the right of the barrier indi-

*Notice that the uncertainty in the position is very large—the electron could be just about anywhere to the left of the forbidden region.

cates that the probability of finding the electron on the right is much lower than the probability of finding it on the left. The probability of tunneling decreases exponentially as the barrier thickness increases—if you double the thickness, the probability is much less than half of the original probability. On the other hand, as the energy of the incoming electrons increases, they penetrate farther into the forbidden region, and the probability of one making it all the way through increases.

"So the electrons just drill holes through the barrier?"

"No, they pass through it as if it weren't there at all. They don't have enough energy to punch through."

"But how do you know that?"

"Well, the electrons show up on the far side of the barrier with exactly the same energy as before they hit it. If they were boring little holes through the barrier, they would lose some energy in the process, and we'd be able to detect that."

"Maybe they're just really tiny holes?"

"No, we can look at that with a scanning probe microscope, and there aren't holes."

"What's a scanning probe microscope?"

"That's an excellent and very convenient question . . ."

FEELING SINGLE ATOMS: SCANNING TUNNELING MICROSCOPY

Tunneling has a more direct technological application than most of the other weird quantum phenomena we've discussed. Tunneling is the basis for a device called a scanning tunneling microscope (STM), which uses electron tunneling to make images of objects as small as a single atom. The STM was invented in 1981 at IBM Zurich, and has become an essential tool for people studying the atomic structure of solid materials. Its inventors, Gerd Binnig and Heinrich Rohrer, won the Nobel Prize in Physics in 1986.

An STM consists of a sample of electrically conducting material and a very sharp metal tip brought within a few nanometers of the surface of the sample. The tip is held at a slightly different voltage than the sample, so electrons in the tip want to move from the tip into the sample. The electrons can't flow directly from the tip into the sample, though, because the small gap between the tip and the sample acts as a barrier preventing the movement of electrons.*

If the gap between the tip and the sample is small enough, though—a nanometer or so—there's some chance that electrons will tunnel from the tip to the sample. That produces a small current, which can be measured. The tunneling probability (and thus the current) increases dramatically as the tip gets closer to the surface, so changes in the current can be used to detect tiny changes in the distance between the two—changes smaller than the diameter of a single atom.

Making an image with an STM is like running your finger across a surface, and feeling the bumps and scratches. You scan the tip back and forth over the surface of the sample, keeping the height of the tip constant, and as you move the tip, you monitor the current flowing between the tip and the sample. The current increases whenever there's a small bump sticking up from the surface making it easier for electrons to tunnel across, and decreases whenever there's a small dip in the surface. If you take a large number of height measurements at points on a grid, you can put them together to create an image of the individual atoms making up the surface of your sample.

Not only can you see single atoms, but if you bring the tip

*A potential energy barrier doesn't have to be a solid physical object. An air gap will do just fine, which is why you can't make a lightbulb light up by just holding it close to the socket.

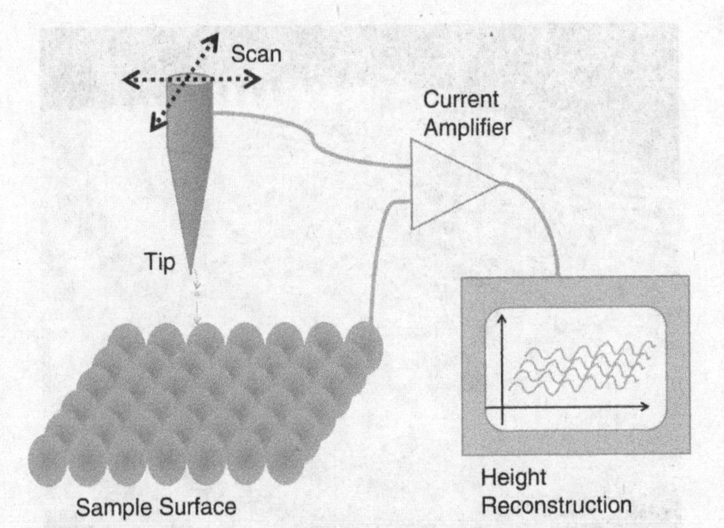

A schematic of a scanning tunneling microscope. A sharp tip is positioned close to the surface of a material, and moved back and forth in a regular pattern. Electrons from the tip will tunnel across the gap between the tip and the surface, producing a small electric current that is amplified and measured. The amount of current depends very sensitively on the distance between the tip and the surface, allowing a reconstruction of the surface sensitive enough to detect single atoms.

into direct contact with the surface, you can push individual atoms around. Scientists have used this ability to make a number of incredible structures, such as the oval-shaped "corral" shown in the picture on the next page, made at IBM's Almaden research laboratory. The bumps making up the "corral" are individual iron atoms on a copper surface, which have been dragged into place by the STM. Such structures can be used to study the quantum behavior of electrons inside the "corral," which accounts for the wavy features seen on the copper surface.

Scanning tunneling microscopes have revolutionized the study of solids and surfaces, and the technology may lead to new manufacturing techniques for tiny devices. Other scientists

Iron atoms on a copper surface, arranged in a "corral" pattern using an STM. The wave pattern inside the corral is due to the wave nature of electrons on the copper surface. Image courtesy of IBM.

have used STMs to study and manipulate individual strands of DNA, providing a more detailed understanding of the behavior of genetic material and possibly new drugs or treatments for genetic damage. All of this is made possible by the underlying wave nature of matter.

"That's nice and all, but I'm not interested in microscopic bunnies. What has quantum tunneling ever done for me?"

"Well, for one thing, you wouldn't be able to enjoy a nice sunny day if not for tunneling."

"What do you mean?"

"Well, the Sun shines because of fusion reactions in the core, right?"

"Everybody knows that. Even the beagle down the street knows that, and that dog is really dumb."

"Yes. Well. Anyway, fusion works by sticking protons together to make helium from hydrogen. Because protons are positively charged, they repel one another, and that repulsion sets up a barrier. And as hot as the Sun is, the protons in the Sun still don't have enough energy to get over that barrier directly."

"So they tunnel through?"

"Exactly. The probability of any given proton tunneling through the barrier is pretty low, but there are lots and lots of protons in the Sun, and enough of them do tunnel through to keep the reaction going. So it's really tunneling that lets the Sun shine."

"Hmm. I guess that is pretty cool."

"I'm so glad you approve . . ."

"Can we go play fetch, now?"

CHAPTER 7

Spooky Barking at a Distance: Quantum Entanglement

Emmy is napping in the living room, but wakes up as I pass through. She stretches hugely, then follows me into the kitchen looking pleased with herself. "I'm going to measure a bunny," she announces.

"Beg pardon?" She's always making these weird announcements.

"I've figured out how to measure both the position and the momentum of a bunny."

"You have, have you? How are you going to do that?"

"I'm going to put a big grid of lines in the backyard, and then when the bunny is right on top of a grid mark, all I have to do is measure how fast it's going." She wags her tail proudly. "Uncertainty, unschmertainty."

"Uh-huh. And how are you going to measure when the bunny is right on a grid mark?"

"What do you mean? I'm just going to look."

"Sure, which means you'll see the bunny, and the bunny will see you, and then it will change its velocity to run away."

"Oh." Her tail droops. "I didn't think of that."

"Look, we've been through this. There's no way around the uncertainty principle. Really smart humans have tried to find a way around it, and it can't be done. Einstein spent years arguing about it with Niels Bohr."

"Did he come up with anything?"

"He tried lots of different arguments, but none of them actually worked. He even had a really clever argument that quantum mechanics was incomplete, involving two entangled particles, prepared so that their states are correlated."

"Correlated how?"

"Well, let's say I have two treats in my hand—stop drooling, it's a thought experiment—and one of them is steak, and the other is chicken."

"I like steak. I like chicken." She's drooling all over the floor.

"Yes, I know. Thought experiment, remember?" I grab some paper towels to mop up the floor. "Now, imagine I throw these two treats in opposite directions, one to you, and one to some other dog."

"Don't do that. Other dogs don't deserve treats."

"It's a hypothetical, try to keep up. Now, if you got the steak treat, you would know immediately that the other dog got the chicken treat. And—why are you looking all sad?"

"I like hypothetical chicken treats."

"You got a hypothetical steak treat."

"Oooh! I like hypothetical steak."

"The point is, by measuring what sort of treat *you* got, you know what the other treat is, without ever measuring it."

"Yeah, so? What's weird about that?"

"Well, in the quantum version, the state of the particles is indeterminate until one of them is measured. When I throw the treats, until you get one and find out whether it's steak or chicken, it's not either. In some sense, it's both."

"Chickensteak! Steakchicken! Sticken!"

"You're ridiculous. Anyway, Einstein thought this was a problem, and that the fact that you could predict the state of one particle by measuring the other particle meant that both of them had to have definite states the whole time."

"That makes sense."

"In a classical world, sure. Einstein's argument fails, though, because he's assumed what's called 'locality'—that measuring one particle does not affect the other. In fact, measuring the state of one determines the state of the other, absolutely and instantaneously."

She looks really bothered by this. "I don't like that idea. Wouldn't that require a message to travel from one treat to the other?"

"That's what bothered Einstein, and he called it *spukhafte Fernwirkung*."

"'Spooky action at a distance'?" she translates.

"Since when do you know German?"

"Dude, look at me." She turns sideways for a second, showing off her black and tan coloring and pointed nose. "*German shepherd*, remember?"

"Of course, how silly of me. Anyway, yes, this bothered Einstein because information cannot pass between separated objects faster than the speed of light. But quantum mechanics is *nonlocal*, and the entangled particles act like a single object. A guy named John Bell showed that it's possible to put limits on what you can measure in theories where the particles have definite states, and showed that those limits are different than the limits for entangled quantum particles. People have done the experiments and found that the quantum theory is right. The state of the particles really is indeterminate until they're measured."

"So Einstein was wrong?"

"About this, yes. And generally, about the basis of quantum theory."

"But he was really smart, wasn't he?"

"Yes. Einstein was arguably smarter than Bohr. Bohr won all their debates, though, because he had the advantage of being right." I bend over to scratch behind her ears. "You're pretty smart, but you're no Einstein."

"I'm, like, the canine Einstein, though, right?"

"Sure. As far as I know, you're the Einstein of the dog world."

"Can I have some steak, then? Or chicken?"

"Maybe." I grab a treat out of the jar on the counter. "You'll find out when you measure it." I throw the treat out the back door, and she goes bounding after it.

"Oooh! Indeterminate treats!"

Everything we have talked about so far has been a one-particle phenomenon. Most of the experiments need to be repeated many times to see the effects, using different individual particles prepared the same way, but at a fundamental level, all the interference, diffraction, and measurement effects we've talked about work with one particle at a time. Each particle in an interference experiment can be thought of as interfering with itself, and measurement phenomena like the quantum Zeno effect involve the state of a single particle.*

Of course, the world we live in involves a great many particles, so we need to look at what happens when we apply quantum physics to systems involving more than one particle. When we do, it's no surprise that we find some weird things going on, starting with the idea of "entangled states."

In this chapter, we'll look at the idea of "entangled" particles, whose states are correlated so that measuring one particle determines the exact state of the other. Entangled particles are the basis for the most serious challenge Einstein mounted against

*The process of decoherence (described in chapter 4) involves the interaction of a single quantum particle with a much larger environment, but we care only about the state of the single particle.

quantum theory, known as the **Einstein, Podolsky, and Rosen (EPR) paradox.** We'll talk about John Bell's famous theorem resolving the EPR paradox, and its disturbing implications for the commonsense view of reality. Finally, we'll talk about the experiments that prove Bell's theorem, and show the lengths that physicists go to in challenging new ideas.

SLEEPING DOGS LET EACH OTHER LIE: ENTANGLEMENT AND CORRELATIONS

Entanglement is fundamentally about correlations between the states of two objects. To illustrate the idea, let's think about two dogs—we'll use my parents' Labrador retriever, RD, and my in-laws' Boston terrier, Truman—who can each be in one of two states: "awake" or "asleep." If the dogs are completely separate from each other, there are four states we could find our two-dog system in: we can find both dogs awake, both dogs asleep, Truman awake while RD is asleep, or Truman asleep while RD is awake.

If we bring the two dogs together and allow them to interact, though, a correlation develops between the state of the two dogs. If Truman is asleep while RD is awake, RD will wake Truman up to play, and vice versa. You will either find both dogs awake or both dogs asleep, but never one awake and the other asleep. We go from four possible states to only two.

Moreover, this correlation allows us to know the state of one of the dogs without measuring it. If Truman is awake, we know that RD must be awake, and if Truman is asleep, we know that RD must be asleep. We can look at RD if we want, but we'll just confirm what we already know. Measuring the state of one of the two dogs immediately and absolutely tells us the state of the other dog.

IS QUANTUM MECHANICS INCOMPLETE? THE EPR ARGUMENT

What does this have to do with Einstein? Einstein was a strong believer in a deterministic universe, in which we can always trace a clear path from cause to effect. He had major philosophical problems with quantum mechanics. In particular, he was bothered by the idea that properties of quantum particles are undefined until they are measured, and then take on random values.

From the late 1920s through the mid-1930s, Einstein had a series of arguments with Niels Bohr, who was also philosophically inclined* but was a champion of the quantum theory. Einstein first attacked the idea of uncertainty with a number of different ingenious thought experiments that would perform measurements forbidden by the uncertainty principle—measuring both the position and momentum of an electron, for example. Every time he did, Bohr found a semiclassical counterargument showing that Einstein's proposed experiment had some hidden flaw.†

In the early 1930s, Einstein reconciled himself to uncertainty, but he remained troubled by quantum theory, and found a new problem to attack. He argued that the existing quantum theory did not contain all the information needed to describe a particle's properties. In a 1935 paper titled "Can Quantum-

*Werner Heisenberg, who developed the uncertainty principle while working with Bohr, once described Bohr as "primarily a philosopher, not a physicist."

†In almost all of those cases, Bohr's argument depended on the effect of measurement on the system. Something in the process by which Einstein proposed to measure the position would cause a change in the momentum (as in the case of the Heisenberg microscope thought experiment discussed in chapter 2 [page 38]), or vice versa. Measuring the system requires an interaction, and that interaction changes the state of the system in a way that introduces some uncertainty in the quantities being measured.

Mechanical Description of Physical Reality Be Considered Complete?" Einstein and colleagues Boris Podolsky and Nathan Rosen presented an ingenious argument for this claim, using the idea of an entangled state. They proposed an experiment to demonstrate this supposed incompleteness by entangling the states of two particles, and then separating them so that they no longer interact (but their states do not change). You can then measure the two in separate experiments that have no possible influence on each other, and see what happens.

In the EPR scheme, measuring the position of one of the two particles (Particle A) allows you to predict the position of the other (Particle B) with absolute certainty. At the same time, if you measured the momentum of Particle B, you would know with certainty the momentum of Particle A. According to Einstein, Podolsky, and Rosen, since there's no way for measurements of Particle A to affect the outcome of measurements of Particle B, or vice versa, both the position and the momentum of each particle must have definite values the whole time. This suggests that quantum mechanics is incomplete: the information needed to describe the precise state of the particles exists, but is not captured by quantum theory.

"That's just what I was saying!"

"What was?"

"A bunny does so have a definite position and momentum. All that uncertainty business was just you being all confusing and stuff."

"It sounds like a convincing argument, but if you remember, I also said it was wrong. It's brilliantly wrong, but there's still a flaw in one of their assumptions, namely the idea that it's impossible for a measurement of one particle to affect the outcome of the state of the other particle."

"Oh, yeah? Prove it."

"I'll get there. Just give me a minute . . ."

"DON'T KNOW" VS. "CAN'T KNOW": LOCAL HIDDEN VARIABLES

Bohr's initial response to the EPR argument was rushed and nearly incomprehensible.* He refined this later, but he was never able to come up with a convincing semiclassical counterargument, in the way that he had in all his other debates with Einstein. The reason is simple: there is no such argument. Quantum mechanics is a "nonlocal" theory, meaning that measurements separated by a large distance can affect one another in ways that wouldn't be allowed by classical physics.

The sort of theory preferred by Einstein, Podolsky, and Rosen is called a **local hidden variable** (LHV) theory, after the underlying assumptions that make up the model. "Hidden variable" means that all quantities that might be measured have definite values, but those values are not known to the people doing the experiment. "Local" means that measurements and interactions at one point in space can only instantaneously affect things in the immediate neighborhood of that point. Long-distance interactions are possible, but those interactions must take some time to be communicated from one place to another, at a speed less than or equal to the speed of light.†

Locality is so central to classical physics that it may seem too obvious to challenge. Locality says that some time must pass

*Bohr was somewhat famous for the opacity of his writing, but he outdid himself in this case. The crucial paragraph of his paper refers to the quantum connection between distant objects as "*an influence on the very conditions which define the possible types of predictions regarding the future behavior of the system*" (italics in original), and declares that the quantum view "may be characterized as a rational utilization of all possibilities of unambiguous interpretation of measurement, compatible with the finite and uncontrollable interaction between the objects and the measuring instruments of quantum theory."

†This light-speed limit is one of the main consequences of Einstein's theory of relativity, and thus very important to his conception of physics.

between causes and effects. When a human calls to a dog out in the yard, the dog won't come running until enough time has passed for the sound of the call to travel from the human to the dog.* Nothing the human does can have any influence on the dog's actions before that time.

Locality is what makes the EPR argument a paradox. Nothing in the proposed experiment limits the time between the two measurements. You can keep Particle A at Princeton, and send Particle B to Copenhagen, and agree to measure the position of A and the momentum of B at, say, one nanosecond past noon, Eastern Standard Time. There is no possible way for any message to travel from Princeton to Copenhagen in time to influence the outcome of the second measurement. Hence, assuming locality is true, the two measurements are completely independent of each other, and each must reflect some underlying reality.

As obvious as the assumption of locality seems, this is exactly the point where the argument fails. Quantum mechanics is a *nonlocal* theory, and a measurement made on one of two entangled objects will affect measurements made on the other instantaneously, no matter how far apart the two are. A measurement in Princeton *can* determine the result of a measurement in Copenhagen, provided the objects being measured are entangled.

Because quantum mechanics is nonlocal, the state of two entangled particles remains indeterminate until one of the two is measured. Not only do you not know the state of the particles, you *can't* know it. In terms of our dog example (page 143), until somebody measures the state of one of the two dogs, both dogs are simultaneously asleep and awake—the wavefunction for the system has a part corresponding to "Truman asleep and RD asleep" and a part corresponding to "Truman awake and RD awake," but neither dog is definitely asleep or awake. The dogs exist in a superposition, like a friendlier version of Schrödinger's cat.

*Or even longer, depending on what the dog is doing when called.

The state of a given dog takes on a definite value only when it is measured, and when that happens, the state of the other dog is simultaneously determined. The instant that you measure one, you determine the state of both, no matter where they are. If Truman is awake, so is RD, and if Truman is asleep, so is RD. If you take them into different rooms before measuring their states, you'll still find them correlated, despite the fact that measuring Truman's condition does not directly affect RD, and no information passes between them. The two separated dogs are a single quantum system, and a measurement of any part of that system affects the whole.

Nonlocality prevents the EPR experiment from being able to circumvent the uncertainty principle. A measurement of Particle A *does* perturb the state of Particle B, exactly as if the measurement had been made on Particle B. This holds true no matter how carefully the two are separated before the measurement—the entangled particles are a single, nonlocal quantum system.

Nonlocality presents a philosophical challenge to the basis of classical science as profound and disturbing as the issues of probability and measurement discussed in chapters 3 and 4. The instantaneous projection of the entangled objects onto definite states* is a conclusion that we're forced to by quantum theory, and there's nothing like it in classical mechanics.

With the EPR paper, quantum physics reached a philosophical impasse. Supporters of Bohr's orthodox quantum theory were unconvinced by the EPR argument, but could not present a compelling counterargument. Meanwhile, people like Einstein who

*If you prefer the Copenhagen view, this projection involves a real collapse of the wavefunction into a single state. If you prefer many-worlds, the apparent projection onto a single state comes because we perceive only a single branch of the wavefunction. In either case, the resulting correlation is the same, and the effect is instantaneous.

were bothered by the implications of quantum theory pointed to the EPR argument as suggesting some deeper theory that would make sense of this weird and unpleasant quantum business. More people took Bohr's side than Einstein's, because quantum theory provided such accurate predictions of atomic properties, but neither side could think of a definitive experiment.

SETTLING THE DEBATE: BELL'S THEOREM

This impasse lasted for almost thirty years, until the Irish physicist John Bell came up with a way to distinguish between the predictions of quantum theory and those of the local hidden variable models preferred by Einstein. Bell realized that LHV theories have definite particle states and only local interactions, and are thus limited in ways that quantum theory is not. He proved a mathematical theorem stating that entangled quantum particles have their states correlated in ways that no possible local hidden variable theory can match. These correlations can be measured experimentally; a measurement showing correlations beyond the LHV limits would conclusively prove that Bohr was right, and Einstein was wrong.

Bell's theorem is critical to the modern understanding of quantum mechanics, so it's worth exploring in some detail. It can't be demonstrated with dogs, but it's not too hard to do using the polarized photons we talked about in chapter 3 (page 65). To be concrete, let's think about two photons whose polarizations are the same—if one is measured to be horizontal, the other is also horizontal; if one is measured at a 45° angle, the other is at the same 45° angle. Then we look at three different possible measurements.

The traditional arrangement calls for two experimenters—they're usually called "Alice" and "Bob," but we'll stick with "Truman" and "RD," because they're good dogs—to each receive one of the two photons. Truman and RD are each given a polarizer and a photon detector, which combine to make detectors that register either a "1" or a "0," depending on whether the photon makes it through the polarizer or not. For example, if the polarizer is set to vertical, a vertically polarized photon will be transmitted and give a "1," while a horizontally polarized photon will be blocked, and give a "0." If the polarizer is set at 45° to the vertical, a vertically polarized photon has a 50% chance of making it through and being recorded as a "1," otherwise it will be blocked and recorded as a "0."

The experiment is simple: each dog sets his polarizer at one of three angles, a, b, or c. He then records the detector reading ("0" or "1") for one photon. Then they change the detector settings, and do it again. After repeating this over and over, they will have tried all the possible combinations of detector settings many times, and then they compare their results.

When they compare results, they'll notice two things. When their polarizers are set to the same angle, they'll see that both get the same answer (either "1" or "0"), every time. They'll also see that no matter what angle they choose, they get equal num-

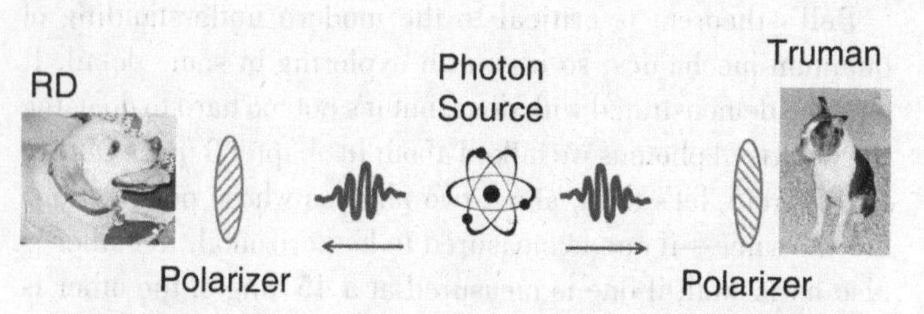

Schematic of a measurement to test Bell's theorem. Truman and RD each take one photon from an entangled photon source, and measure its polarization along one of three angles using a polarizing filter and a photon detector. They can distinguish between quantum mechanics and a local hidden variable theory by measuring how often they detect the same thing when their polarizers are at different angles.

bers of "0" and "1" results—if they repeat the experiment 1,000 times at a given angle, they will get 500 "0"s and 500 "1"s. These two observations are true whether they're dealing with a quantum entangled state, or a state governed by an LHV theory.

"Wait, shouldn't it depend on the angles?"

"What angles?"

"Your a, b, and c angles. Why do they get equal numbers of '0's and '1's? Shouldn't the measurement results depend on which angle they choose? Like, if they have their polarizers set vertically, they always detect a '1'?"

"No, the states we're dealing with are states of indeterminate polarization. In the quantum picture, the polarization is undefined, while in the LHV picture, it's equally likely to be either horizontal or vertical."

"Doesn't that mean they're at 45°? Then shouldn't they get '1' every time when they put the polarizers at 45°?"

"No, they get the same result at 45°. The photons are equally likely to be 45° counterclockwise from vertical, or 45° clockwise, or any other angle. It really doesn't matter what angles they choose for a, b, and c—even 'vertical' and 'horizontal' are kind of arbitrary."

"No they're not."

"Yes they are. When I say something weird, and you look at me sideways—like you're doing right now—that changes what 'vertical' looks like, right?"

"I guess. Everything looks different from an angle, and sometimes weird human stuff makes more sense."

"It's the same thing here. The angles they set for the polarizers determine what '0' and '1' will mean, in the same way that tilting your head changes your perception of 'horizontal' and 'vertical.' They still have an equal chance of getting either result. What you see depends on what you're looking for. To go back to the treat analogy from page 140, it's like a treat where if you're looking for 'meat,' you get either steak or chicken, but if you're looking for 'not meat,' you get either peanut butter or cheese."

"Oooh! Those treats sound good. You should buy me some of those."

"I don't think they have them at the pet store, but I'll look."

To test Bell's theorem, we ask how often they get the *same* answer with their detectors at *different* settings. That is, how many times did Truman record a "0" with the detector in position "a," while RD got a "0" in position "c," or Truman a "1" in position "b" and RD a "1" in position "a," and so on. The probability of both dogs getting the same result with different detector settings is very different for LHV theories and quantum mechanics.

THE EPR OPTION: LOCAL HIDDEN VARIABLE PREDICTION

The key to Bell's theorem is that all the predictions of a local hidden variable theory can be written down in advance, so let's do that. Each photon has a well-defined state, and we can represent that state by a set of three numbers, each giving the definite outcome of a measurement in polarizer position a, b, or c. The two-photon system offers a total of eight possible states, which we can represent in a table:

	Truman			RD		
State	**a**	**b**	**c**	**a**	**b**	**c**
1	1	1	1	1	1	1
2	1	1	0	1	1	0
3	1	0	1	1	0	1
4	1	0	0	1	0	0
5	0	1	1	0	1	1

	Truman			RD		
State	a	b	c	a	b	c
6	0	1	0	0	1	0
7	0	0	1	0	0	1
8	0	0	0	0	0	0

To test Bell's theorem, we need the probability of both dogs getting the *same* answer with *different* detector settings. Looking at the table, we see that no matter what angles we pick, four of the eight possible states give the same answer. For example, if Truman sets his detector to position a, and RD sets his to b, states 1 and 2 will give them each a "1" and states 7 and 8 will give them each a "0." If Truman chooses c and RD chooses a, the four states giving the same answer are 1, 3, 6, and 8, and so on.

We're not stuck with exactly 50% probability of getting the same answer for different settings, though. We're free to adjust the probability of the photons being in a particular state—say, making state 1 more likely, and state 6 less likely—though any change we make has to end up with equal probabilities of finding "0" or "1" for each detector setting.

If we play around with the probabilities of the individual states, we find that we can cover a limited range of possible probabilities. We can make the maximum probability of both dogs getting the same result 100%, but the minimum probability is 33%, not 0%. No matter what we do, we can never make the probability lower than 33%.*

*We get the maximum value of 100% if the system has a 50% chance of being in state 1 and a 50% chance of being in state 8. We get the minimum value of 33% by never letting the system be in state 1 or state 8, and making the other six states equally likely. If you look at states 2 through 7, you'll see that no matter what two different angles you choose, there are always two states that give you the same answer for both detectors.

Notice that we haven't said anything about what causes those states, or how they are chosen. We don't need to—the mere fact that we can write down the limited number of possible results places restrictions on the experiment. No model in which the two photons have well-defined states when they leave the source can give a probability of less than 33% for the two measurements to give the same outcome. The probability must be less than or equal to 100%, and greater than or equal to 33%.* Similar limits hold true for any LHV theory you can dream up.

THE BOHR OPTION: QUANTUM MECHANICAL PREDICTION

To prove quantum mechanics correct, then, we need to find some detector angles for which the probability of both dogs getting the same answer with different settings is less than 33%. Bell showed that this can be done, thanks to entanglement: measuring the polarization of one of the two photons instantaneously determines the polarization of the other.

In the quantum picture, the state of the two photons is indeterminate until the instant when one of the two is measured, when it has a 50% chance of ending up as a 0 or 1. At that instant, the polarization of the second photon is set to the same angle as the first, whatever that is. If the first photon passed through a vertical polarizer, recording a "1," the second photon is now vertically polarized. If the first photon was blocked by the vertical polarizer, recording a "0," the second photon is now horizontally polarized. The possible outcomes of the second measurement are then determined by the first polarizer angle.

To prove Bell's theorem, let's imagine Truman sets his detec-

*As a result the predictions of Bell's theorem are often called "Bell inequalities."

tor to vertical polarization (which we'll call "a"). RD sets his detector to either 60° clockwise from vertical ("b"), or 60° counterclockwise from vertical ("c"). What are the possible ways to get the same answer for both dogs when they have different polarizer settings?

Well, half of the time, Truman will detect a "1" with his detector, which means that we want the probability of RD also getting a "1."* Since Truman's polarizer is vertical, the entangled photon hitting RD's detector is also vertically polarized. If his detector is set to position "b," then the angle between the vertical photon and RD's polarizer is 60°, and the probability of the photon passing through the polarizer is 25%. The same holds for position "c," which is 60° from "a" in the other direction.

The other half of the time, Truman measures a "0," and both entangled photons are horizontal. RD's photon again has a 25% chance of being blocked and giving a "0,"† for either angle.

No matter what value Truman measures, then, quantum theory tells us that there is only a 25% chance that RD will get the same value with his detector at a different polarizer setting. This directly contradicts the prediction of the local hidden variable theory, which gave a minimum chance of 33%. Only one in four of RD's measurements is the same as Truman's, where LHV says that *at least* one in three should be the same.

You might think that the two theories should give the same results, because they're describing the same system, in the same way that the different interpretations of quantum mechanics all give the same predictions. That's what most physicists thought,

*We're assuming that Truman's photon is measured first, for the sake of clarity. The result is the same if we assume RD's photon is the first one measured.

†The horizontal photon has a 75% chance of passing through the polarizer to the detector in either position. A "0" to match Truman's result happens only when RD's photon is blocked, a 25% probability.

until Bell showed otherwise. The core assumptions of the local hidden variable theories mean that they are subject to strict limits—you can write down a table like the one above showing all possible results. Quantum theories do not have the same limitations, so a clever experiment can distinguish between them.*

The results are different because quantum mechanics is nonlocal—the polarization of RD's photon is not set in advance, but is determined by the outcome of Truman's measurement. The probability of getting the same result with different settings is lower because the two measurements affect each other, no matter how far apart they are, or when they're made. Einstein called this "spooky," and it's hard to argue with him.

"Can't you just make a better theory?"

"What kind of better theory?"

"A better hidden variable theory. That matches the predictions better."

"That's the whole point. Bell didn't look at a *particular* theory—what he showed is that there's *no possible* local hidden variable theory that can reproduce all the predictions of quantum mechanics. If the two measurements are independent of each other, there's no way to arrange things so that the measurements show the same correlation that you see with quantum mechanics."

"So make the measurements depend on each other."

"That works, but that isn't a *local* hidden variable theory anymore. In fact, David Bohm worked out a version of quantum mechanics that uses *non*local hidden variables, and reproduces

*This raises the question of whether a sufficiently clever experiment might distinguish between, say, the Copenhagen interpretation and the many-worlds interpretation. This is a much harder problem than distinguishing between quantum and LHV theories. Some future John Bell may yet come along and find the right test, but no one has managed yet.

all the predictions of quantum theory using particles with definite positions and velocities."

"That sounds nice. Why don't people use that?"

"Well, Bohm's theory introduces an extra 'quantum potential,' a function that extends through the entire universe and changes instantaneously when you change some property of the experiment. It's a really weird object, and it's a headache when doing calculations. It's also easier to extend regular quantum mechanics to be compatible with relativity, in what's known as **quantum field theory.**"

"It's not wrong, though?"

"No, it predicts the same things as regular quantum theory. You can look at it as an extreme version of a quantum interpretation, like the Copenhagen interpretation or many-worlds pictures that we talked about earlier. It adds a little more math to the theory, but doesn't predict anything different in practical terms."

"Hmmm."

"The important thing for this discussion is that Bohm's theory is nonlocal, which is what the EPR paradox and Bell's theorem are really about. From those, we know that quantum theory can't be a strictly local theory, where measurements in two different places have no effect on each other."

"That still bugs me. How do we know that that's really true?"

"I'm glad you asked that . . ."

This example is a specific demonstration of Bell's theorem, but it captures the flavor of the general theorem. What Bell showed is that there are limits on what can be achieved with LHV theories in general, and that under certain conditions, quantum mechanics will exceed those limits. A clever experiment can determine once and for all whether quantum mechanics is right, or whether it could be replaced by a local hidden variable theory as Einstein hoped.

LABORATORY TESTS AND LOOPHOLES: THE ASPECT EXPERIMENTS

Bell published his famous theorem in 1964. In 1981 and 1982, the French physicist Alain Aspect and colleagues tested Bell's prediction with a series of three experiments that are generally considered to conclusively rule out local hidden variable theories.* They needed all three experiments to close a series of "loopholes," gaps in their results that some local hidden variable models might slip through.

We'll describe all three experiments here, because they're outstanding examples of the art of experimental physics. More than that, though, they demonstrate the lengths you need to go to if you want to convince physicists of something. You need to answer not only the obvious objections, but also objections that are improbable enough to seem a little ridiculous.

The first experiment, published in 1981, was essentially the same as our thought experiment with Truman and RD. Aspect's group made calcium atoms emit two photons within a few nanoseconds of each other, heading in opposite directions. These photons are guaranteed to have the same polarization—it's equally likely to be either horizontal or vertical (or any other pair of angles), but if one photon is horizontal, the other must also be horizontal. This is exactly the entangled state you need in order to test Bell's theorem.

In the first experiment, they placed two detectors on opposite sides of their entangled photon source, with a polarizer in front of each detector. The polarizers were set to various different angles, and they measured the number of times they counted a

*John Clauser and a couple of other people had done earlier tests, but the Aspect (pronounced "As-PAY") experiments had better precision, and so are regarded as the definitive tests.

The first Aspect experiment. An excited calcium atom emits two photons with entangled polarizations. Each photon heads toward a single detector with a polarizing filter in front of it, set to an appropriate angle.

photon at both detectors—that is, both detectors reading "1," in terms of our example above.

Physicists like to deal with numbers, and for the specific configuration they used, a local hidden variable treatment predicts that their results should boil down to a number between -1 and 0. When they did the experiment, they measured a value of 0.126, with an uncertainty of plus or minus 0.014.* The difference between the maximum LHV value and their measurement is nine times larger than the uncertainty in the measurement, meaning that there's a one in 10^{36} probability of this happening by chance.†

So, that's the end of LHV theories, right? It looks just like our imaginary experiment above, and that's an astonishingly small probability of this happening by accident. Why did they need to do a second experiment, let alone a third?

Unfortunately, there's a loophole in their result that allows some LHV theories to survive. In our thought experiment, we imagined Truman and RD with photon detectors that were abso-

*This uncertainty is a technical limitation based on the details of their experiment, and not anything to do with the Heisenberg uncertainty principle.

†10^{36} is a billion billion billion billion, a number so large that it might have made even Carl "Billions and Billions" Sagan blink.

lutely perfect, because they're very good dogs. Aspect and his coworkers are only human, though, and so were stuck using detectors with limited efficiency. On rare occasions, a detector would fail to record a photon that was really there.

This is a problem, because their experiment recorded a "0" when they expected a photon and didn't see one—they assumed that those photons were blocked by the polarizers. But because their detectors sometimes failed to detect photons, it's conceivable that the first Aspect experiment just *looked* like it violated the LHV prediction. If some of their "0"s really should've been "1"s, that could confuse their results.

For LHV theories to slip through this loophole the universe would need to be somewhat perverse, but it's possible, so they did a second experiment, published in 1982, using two detectors for each photon.

They closed the detector efficiency loophole by directly detecting *both* possible polarizations, and only counting experiments where they detected one photon on each side of the apparatus. They replaced the polarizers with polarizing beam splitters that directed each polarization to its own detector. If one of the detectors failed to record a photon, that run of the experiment was discarded.

Their measured value in the second experiment exceeded the

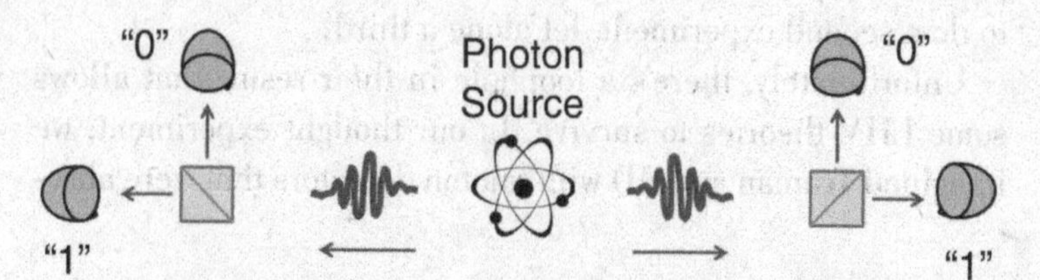

The second Aspect experiment. The entangled photons leave the source and head toward a pair of detectors with a polarizing beam splitter in front of them. These beam splitters direct the "0" polarization to one detector and the "1" polarization to another, ensuring that no photons are missed in the experiment.

LHV limit by an astonishing 40 times the uncertainty, and the odds of that happening by chance are so small it's ridiculous. So, why did they do the third experiment? As impressive as the second experiment was, it still left a loophole, because something could have passed messages between their detectors and their source.

To test Bell's theorem, it needs to be impossible for the measurement at one detector to depend on what happens at the other detector without some faster-than-light interaction. If there's a way to send messages between the detectors at speeds less than that of light, all bets are off. In the first two experiments, they chose the detector settings in advance, and left them set for much longer than it took light to pass between the source and the detector. Something might have communicated the polarizer settings from the detectors to the source, which then sent out photons with definite polarization values chosen to match the quantum predictions. When the experimenters changed the angles, the new values would be sent to the source, which would change the polarizations sent out. Their results *seemed* to prove quantum theory, but they might have been the victims of a cosmic conspiracy.

The third experiment found an ingenious way to close that loophole, as well. Aspect and his colleagues ruled out any possibility of some sort of universal conspiracy mimicking the quantum results by changing their detector settings faster than light could go from the source to the detector.

They replaced the beam splitters with fast optical switches that could direct the photons to one of two detectors, each set for a different polarization. The switches flipped between the detectors every 10 nanoseconds, while it took the photons 40 ns to reach the detector. In effect, which detector a given photon would hit was not decided until *after* the photon had already left the source.

The third experiment's results exceeded the LHV limit by five

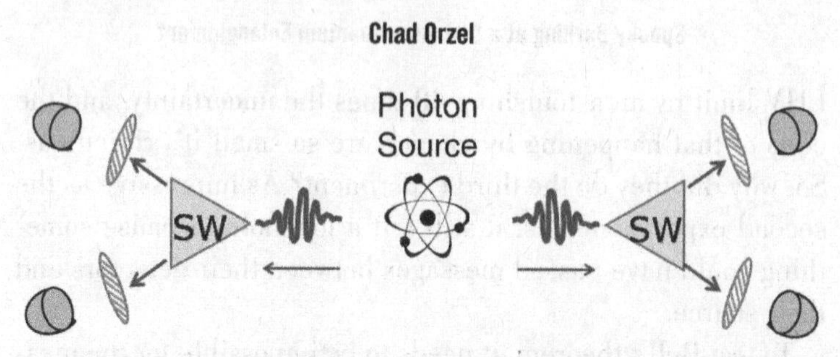

The third Aspect experiment. The two entangled photons leave the source, and head toward fast optical switches that send each photon toward one of two different polarizers, with the choice not being made until after the photons have left the source.

times the uncertainty. The chances of such a result happening by accident were about one in a hundred billion—better than the chances for the other two experiments, but still low enough to be convincing.

Even the third experiment doesn't close every loophole,* but Aspect stopped there, because the experiments were extraordinarily difficult. A number of people have repeated these experiments, using more modern sources of entangled photons,† and a 2008 experiment has even tested Bell's theorem using entangled states of ions instead of photons, but no loophole-free test has been done. As a result, there are still a few people who argue that LHV theories have never been completely ruled out.

These few die-hard theorists aside, the vast majority of physicists agree that the Bell's theorem experiments done by Aspect and company have conclusively shown that quantum mechanics is nonlocal. Our universe cannot be described by any the-

*The third experiment actually reopens the detector efficiency loophole, because they used only one detector for each polarizer.

†One experiment by Paul Kwiat (who was part of the Innsbruck–Los Alamos team doing quantum interrogation experiments in chapter 5) and colleagues at Los Alamos saw an effect a mind-boggling 100 times larger than the uncertainty.

ory in which particles have definite properties at all times, and in which measurements made in one place are not affected by measurements in other places.

Aspect's experiments represent a resounding defeat for the view of the world favored by Einstein and presented in the Einstein, Podolsky, and Rosen paper in 1935. But while the EPR paper is wrong, it's brilliantly wrong, forcing physicists to grapple with the philosophical implications of nonlocality. Exploring the ideas raised in the paper has deepened our understanding of the bizarre nature of our quantum universe. The idea of quantum entanglement exploited in the EPR paper also turns out to allow us to do some amazing things using the nonlocal nature of quantum reality.

"Physicists are really weird."

"Yeah, nonlocality is strange."

"Not that, the loopholes. Do physicists really believe that there are messages being passed back and forth between different bits of their apparatus? What would carry the messages?"

"I'm not sure anybody ever suggested a plausible mechanism, but it really doesn't matter. They could be carried by invisible quantum bunnies, for all the difference it makes."

"Quantum bunnies?"

"Invisible quantum bunnies. Moving at the speed of light. Don't get your hopes up."

"Awww . . ."

"Anyway, the third Aspect experiment pretty much rules out any means of carrying messages between parts of the apparatus, involving bunnies or anything else. The point is, prior to that, it was at least possible in principle for there to be another explanation. And in science, you have to rule out all possible explanations, even the ones that seem really unlikely, if you want to convince anybody of an extraordinary claim."

"Even the ones involving bunnies?"

"Even the ones involving bunnies. And anyway, the idea that distant particles can be correlated in a nonlocal fashion isn't all that much weirder than quantum bunnies would be."

"Good point. So, what's this good for?"

"What do you mean?"

"You dropped a really broad hint in that last paragraph that this entanglement stuff is good for something. What's it good for, sending messages faster than light?"

"No, you can't use it for faster-than-light communication, because the detections are random. There are correlations between particles, but the polarization of each pair will be random. I can't send a message to somebody else using EPR correlations—all I can send is a random string of numbers."

"So what good is it?"

"Well, random strings of numbers can be useful for quantum cryptography, making unbreakable codes. And the idea of entanglement is central to quantum computing, which could solve problems no normal computer can tackle. And there's quantum teleportation, using entanglement to move states from one place to another. There's all sorts of stuff out there, if you look for it."

"Ooh! Teleportation sounds cool! Talk about that."

"Well, that's next . . ."

CHAPTER 8

Beam Me a Bunny: Quantum Teleportation

Emmy trots into my office, looking pleased with herself. This is never a good sign.

"I have a plan!" she announces.

"Really. What sort of plan is this?"

"A plan to get those pesky squirrels." They keep escaping up the trees in the backyard, and she's getting frustrated.

"Is this a better plan than the one where you were going to learn to fly by eating the spilled seed from the bird feeder?"

"That was going to work," she says, indignantly. "And for your information, yes, it's a much better plan than that."

"Well, then, I'm all ears. What's this brilliant plan?"

"Teleportation." She looks smug, and wags her tail vigorously.

"Teleportation?"

"Yep."

"Okay, you're going to have to unpack that a little."

"Well, I figured, the problem is, they can see me coming from the house, and they get to the trees before I do. If I could get between them and the trees, though, I could get them before they get away."

"Okay, I'm with you so far."

"So, I just need to teleport out into the backyard, instead of going through the door." Her whole back end is wagging now.

"Uh-huh. And how, exactly, did you plan to accomplish this feat?"

"Well . . ." The tail slows down, and she does her very best cute-and-pathetic look. "I was hoping you would help me."

"Me?"

"Yeah. I read where some physicists have done quantum teleportation, and you're a physicist, and you're really smart, and you know about quantum, so I was hoping you would help me build a teleporter." She puts her head in my lap. "Pleeeeease? I'm a good dog."

I scratch behind her ears. "You are a good dog, but I really can't help. For one thing, I don't do teleportation experiments in my lab. But even if I did, I wouldn't be able to help you use teleportation to catch squirrels."

"Why not?"

"Well, the existing teleportation experiments all deal with single particles, usually photons. You're made up of probably 10^{26} atoms—a hundred trillion trillion—which is way more than anybody has ever teleported."

"Yeah, but you're really smart. You can just . . . make it bigger."

"I appreciate your confidence, but no. The bigger problem is that the quantum teleportation people do in the real world isn't like the teleportation you see with the transporters on *Star Trek*."

"How so?"

"Well, all that quantum teleportation does is transmit the *state* of a particle from one place to another. If I have an atom here, for example, I can 'teleport' it to the backyard, and end up with an atom there that's in the exact same quantum state as the atom I started with here. At the end of the process, though, I still have the original atom here where it started—it doesn't move from one place to another."

"That's pretty lame. What's the point of that?"

"Well, quantum mechanics won't let you make an exact copy of a state without changing the original state, and quantum states of things like atoms are pretty fragile. If you really needed to get a particular quantum state from one place to another, your best bet might be to teleport it." She looks a little dubious. "You could use it to make a quantum version of the Internet, if you had a couple of quantum computers that you needed to connect together."

"Well, okay. So just teleport my state into the backyard, and I'll use it to catch squirrels."

"Even if I knew how to entangle your state with a whole bunch of photons—which I don't—I would need to have raw material out in the backyard. There would need to be another dog out there, one that looked just like you."

Her tail stops dead. "We don't like those dogs," she says. "Dogs that look just like me. In my yard. We don't like those dogs at all." She looks distressed.

"No, we don't. One of you is all the dog we need." She perks up a bit. "So, you see, teleportation isn't a good plan, after all."

"No, I guess not." She's quiet for a moment, and looks thoughtful. "Well," she says, "I guess it's back to plan A."

"Plan A?"

"Can I have some birdseed?"

"Quantum teleportation" is probably the best-known application of the nonlocal correlations discussed in the previous chapter. The name certainly fires the imagination, conjuring up images of *Star Trek* and other fictional settings in which people, either through fictional science or just plain magic, can instantaneously transport objects from one place to another. The object starts at point A, disappears with a soft **bamf**, and reappears at point B, some distance away.

The high expectations created by science fiction make the reality of **quantum teleportation** seem disappointing. Real quantum teleportation involves only the transfer of a quantum state from one location to another, and not the movement of complete objects. The transfer is also slower than the speed of light, because information needs to be sent from one place to another. This is a great disappointment to dogs hoping to beam themselves out into places where unsuspecting critters are waiting.

Nevertheless, it's a marvelously clever use of quantum theory, tying together several of the topics that we've already talked about. In this chapter, we'll see how indeterminacy and quantum measurement make it difficult to transmit information about quantum states from one place to another. We'll see how the "quantum teleportation" scheme makes ingenious use of nonlocality and entangled states to avoid these problems, and why you might want to.

Quantum teleportation is a complex and subtle subject, probably the most difficult topic discussed in this book. It's also the best example we have of the strangeness and power of quantum physics.

DUPLICATION AT A DISTANCE: CLASSICAL "TELEPORTATION"

We can't teleport in the way envisioned in science fiction and fantasy, but the essence of teleportation is just duplication at a distance—you take an object at one place, and replace it with an exact copy at some other location. By that definition, we do have an approximation of teleportation using classical physics: a fax machine.

If you have a document that you want to send instantly from one place to another—for example, if Truman has just gotten a really nice bone, and wants to taunt RD by sending him a picture of it—you can do this with a fax machine. The machine works by scanning the document, converting it to electronic instructions for creating an identical document, and sending that information over telephone lines to another fax machine at a distant location, which prints a copy. What's transmitted is not the document itself, but rather information about how to make that document.

The operation of a fax machine is different from the fictional idea of teleportation, but the differences are not all that significant. When you fax a document from one place to another, you end up with two copies in different locations, but if you regard this as a problem, you could always attach a shredder to the sender's fax machine to destroy the original. The copy produced by a fax machine isn't perfect, but that's just a matter of the resolution of the scanner and printer, and you can always imagine getting a better scanner and printer. The transmission is limited by the time it takes to transmit the information from one place to another, so it's not perfectly instantaneous, but that's not a major problem for most transactions involving a fax machine.

If you wanted to approximate the fictional ideal of teleportation in a classical world, the best you could do would be to upgrade the concept of the fax machine. Truman would take a bone, and place it in a machine, which would scan the bone to determine the arrangement of atoms and molecules making up the bone. Then he would send this information to RD's "teleportation" machine, which would assemble an identical bone out of materials at hand and present it to him to chew.

NO CLONING ALLOWED: QUANTUM LIMITATIONS

When we turn to quantum teleportation, we're talking about "teleporting" a quantum object. This means not just getting the right physical arrangement of the atoms and molecules making up the object, but also getting all those particles in the right quantum states, including superposition states. Truman could use an upgraded fax machine to send RD a cat in a box, but he would need a quantum teleportation device to send a cat in a box that was 30% alive, 30% dead, and 40% bloody furious. This turns out to be vastly more difficult than the classical analogue, due to the active nature of quantum measurement.

While in theory it is possible to do quantum teleportation with any object, in practice, all of the experiments done to date have used photons, so we'll imagine that Truman is trying to send a single photon of a particular polarization to RD.* As we saw back in chapter 3 (page 65), a polarized photon can be thought of as a superposition of horizontal and vertical polarizations, with some probability of finding either of those two allowed states.

When we describe a photon with a polarization between vertical and horizontal, we write a wavefunction for that photon that is a superposition state: it's a parts vertical, and b parts horizontal:

$$a\,|V\rangle + b\,|H\rangle$$

The numbers a and b tell us the probability of finding vertical or horizontal polarization.† In fact, any object in a superposition state will be described by a wavefunction exactly like this one. If we can find a way to teleport a photon polarization from Truman to RD, we can use the same technique to teleport the state of a cat in a box—it's just a matter of increasing the number of particles involved.

So, Truman has a photon that he wants to send to RD. The classical recipe tells him to simply measure the polarization

*We'll talk about why he might want to do such an odd thing at the end of the chapter.

†Strictly speaking, a^2 is the probability of finding vertical polarization, and b^2 the probability of horizontal polarization and $a^2 + b^2 = 1$. So for a photon at 30° from the vertical, with a 75% chance of passing a vertical polarizer, $a = \sqrt{3}/2$, and b = 1/2.

of the photon, then call RD on the phone, and tell him how to prepare an identical state. But the only way Truman can measure the polarization is if he already knows something about the state, and can set his polarization detector appropriately. For example, if he knows that the photon is either vertical or horizontal, he can send it at a vertically oriented polarizer. If it passes through, he knows that the polarization was vertical, and if it gets absorbed, he knows it was horizontal. He can then send that information to RD, who can prepare a photon in the appropriate state.

Unfortunately, if the polarization is at some intermediate angle—*a* parts vertical and *b* parts horizontal—it's impossible for Truman to make the necessary measurement. The numbers *a* and *b* tell us the probability of the photon passing through a vertical or horizontal polarizer, but there's no way of measuring both *a* and *b* for a *single* photon—either it passes through a polarizer or it doesn't. Even if the photon passes through, the superposition is destroyed and it's left in one of the allowed states.

You can only determine both probabilities by repeating the measurement many times using identically prepared photons. That doesn't help us to transmit the polarization of a *single* photon, though, which is our goal.

This polarization measurement problem is a specific example of the **no-cloning theorem.** William Wootters and Wojchiech Zurek proved in 1982 that it is impossible to make a perfect copy of an unknown quantum state. Unless you already have some idea what the state is, you change the state when you try to measure it, and can never be sure that your copy is faithful. If Truman really needs to send RD a perfect copy of a single photon, without knowing its polarization in advance, he'll need to find a more clever way of doing it.

"Why not just send the photon?"

"Pardon?"

"I mean, it's a photon. They travel places at the speed of light—that's what they *do*. If I had a photon and I wanted to send it to some other dog—which I don't, by the way. Other dogs don't *deserve* my photons. If I did, though, I would just point the photon at the other dog, and let it go."

"Oh. Well, there are a lot of things that can happen to a photon on the way from one place to another that would change the polarization. If you want to be sure that the dog on the other end gets exactly the polarization you started with, teleportation is a sure way of doing that."

"That's a silly thing to want to do, anyway."

"Not really, but you'll have to wait until the end of the chapter to find out why."

A MAGIC COMPASS: CLASSICAL ANALOGUE OF QUANTUM TELEPORTATION

It's hard to find a classical analogue for quantum teleportation, because the issues involved are inherently nonclassical. But we can get a little of the flavor of what's involved by thinking of the photon teleportation process in graphical terms. We can also get a hint of what quantum teleportation will really require.

As we saw in chapter 3 (page 66), we can represent a photon polarization by an arrow indicating the direction of polarization. We can think of the horizontal and vertical components in terms of the number of steps we take in the different directions: you take a steps in the vertical direction, and b steps in the horizontal direction.

In this graphical picture, teleportation is a problem of aligning arrows. Truman has an arrow pointing in some direction, and both dogs will get steak if RD can make his arrow point in the same direction. How do they manage this?

The only way the two dogs can get their arrows aligned is if

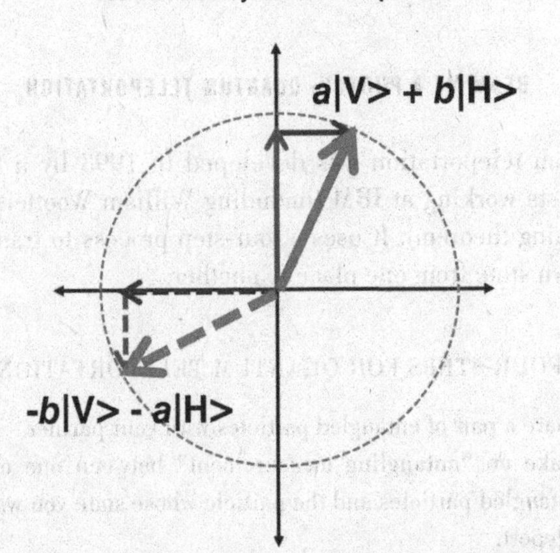

Representing light polarization as a sum of horizontal and vertical components. The larger arrows represent two different photon states, while the smaller arrows are the vertical and horizontal components.

they have some shared reference. If they each have a compass, Truman can compare his arrow to the direction of the compass needle, and tell RD to point his arrow, say, 17° east of due north. The compass provides a reference that they both share, and any scheme for photon teleportation will need a similar reference.

The problem of teleporting a photon is much harder than simply aligning arrows, though, because of the no-cloning theorem. Truman can't measure the direction of his arrow without disturbing it. Somehow, he needs to communicate the direction of his arrow to RD without measuring it. What he needs is a *nonlocal* reference, a kind of magic compass that can communicate a direction to RD's compass without making a measurement. Quantum teleportation is possible because the quantum entanglement that we discussed in chapter 7 provides this kind of nonlocal reference.

BEAM ME A PHOTON: QUANTUM TELEPORTATION

Quantum teleportation was developed in 1993 by a team of physicists working at IBM (including William Wootters of the no-cloning theorem). It uses a four-step process to transfer an unknown state from one place to another:

FOUR STEPS FOR QUANTUM TELEPORTATION

1. Share a pair of entangled particles with your partner.
2. Make an "entangling measurement" between one of the entangled particles and the particle whose state you want to teleport.
3. Send the result of your measurement to your partner by classical means.
4. Tell your partner how to adjust the state of his particle according to the measurement result.

This recipe for teleportation exploits quantum entanglement to generate a copy of an arbitrary state at a distant location through one measurement and a phone call. It uses the active nature of quantum measurement to align one of the two entangled photons with the state to be "teleported." In the process, the second entangled photon is instantly converted to a polarization that depends on the original state. The no-cloning theorem still applies, so the state of the original particle is altered by the measurement, but at the end of the process, the second entangled photon is in the same state as the original photon before "teleportation."

Here's how it works: let's imagine that Truman has a single photon in a particular polarization state, and he wants to get exactly that state to his old friend RD (but he can't just send it straight there). Anticipating that this situation might come up, Truman and RD have previously shared a pair of photons in an entangled state, each taking one. The polarizations of these photons are indeterminate until measured, but they are guaranteed to be opposite each other. So, the two dogs have a total of three photons: Photon 1 is the state that Truman wants to convey to RD (at some randomly chosen angle, described by $a|V\rangle + b|H\rangle$), Photon 2 is Truman's photon from the entangled pair, and Photon 3 is RD's photon from the entangled pair. The teleportation procedure outlined above will allow RD to turn his Photon 3 into an exact copy of Photon 1.

Teleportation works because quantum physics is nonlocal. We saw in chapter 7 that any measurement Truman makes on

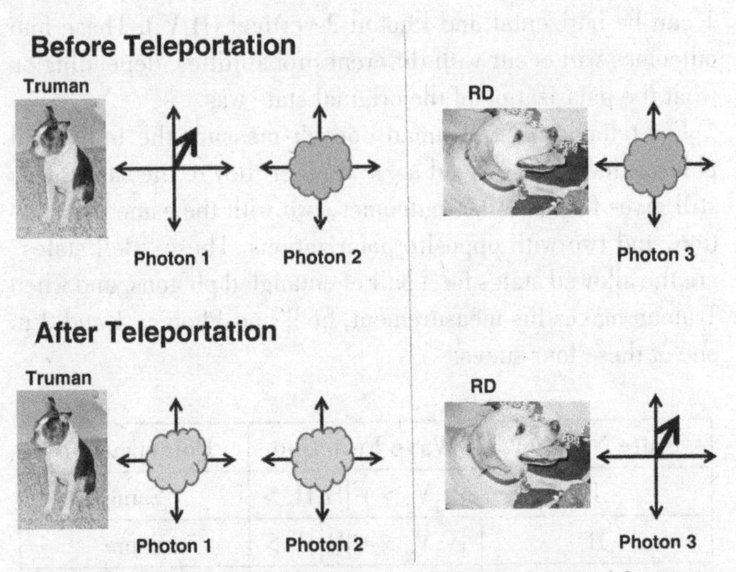

A cartoon version of quantum teleportation. At the beginning of the process, Truman has two photons, Photon 1 in a definite (though unknown) state that he wants to send to RD, and Photon 2 in an indeterminate state that is entangled with RD's Photon 3. After the teleportation procedure is completed, Truman has two photons in an indeterminate state entangled with each other, and RD has a photon whose polarization is identical to the original polarization of Photon 1.

Photon 2 will instantaneously determine the polarization of RD's Photon 3. Of course, it's not as simple as measuring the individual polarizations of Photon 1 and Photon 2—we already saw that that won't work. Instead, what Truman does is to make a *joint* measurement of the two photons together. He measures whether the two polarizations are the same or different—not what they are, just whether they're the same.

If Truman measured the two photons individually, asking whether they're horizontal or vertical, there are four possible outcomes. Both photons can be vertical (we write this as V_1V_2, where the first letter indicates the polarization of Photon 1 and the second that of Photon 2), both can be horizontal (H_1H_2), Photon 1 can be vertical and Photon 2 horizontal (V_1H_2), or Photon 1 can be horizontal and Photon 2 vertical (H_1V_2). These four outcomes will occur with different probabilities, depending on what the polarization of the original state was.

For teleportation, Truman doesn't measure the individual polarizations, but instead asks whether they're the same. This still gives four possible outcomes, two with the same polarization, and two with opposite polarizations. These "Bell states" are the allowed states for a pair of entangled photons, and when Truman makes his measurement, he'll find Photons 1 and 2 in one of these four states:

State Number	Wave Function	Polarizations
I	$\|V_1V_2> + \|H_1H_2>$	Same
II	$\|V_1V_2> - \|H_1H_2>$	Same
III	$\|V_1H_2> + \|H_1V_2>$	Opposite
IV	$\|V_1H_2> - \|H_1V_2>$	Opposite

These states are superpositions of the four possible outcomes from the independent measurements, just like Schrödinger's famous cat is in a superposition of "alive" and "dead."* Each of the polarizations is still indeterminate—if you go on to measure the individual polarization of Photon 1, you are equally likely to get horizontal or vertical. When you do measure Photon 1, you determine the state of Photon 2 to be either the same or opposite, depending on which of the four states you're in.

"Wait a minute—why are there four outcomes? Shouldn't there just be two? What's with the pluses and minuses? Either they're the same, or they're not."

"That's true, but in quantum mechanics, there are two different states where they have the same polarization, State I and State II, and two where they have opposite polarizations, State III and State IV. That gives four states."

"But what's the difference between State I and State II?"

"They're different states, in the same way that |V> + |H> and |V> – |H> are different states for a single photon."

"Wait—they are?"

"Sure. You can see it by thinking of how they add together to give a single polarization at a different angle. You can imagine the |H> as being one step either left or right, and the |V> being one step either up or down. |V> + |H> is then one step up, and one to the right, while |V> – |H> is one step up, and one to the left."

"So, |V> + |H> is 45° to the right of vertical, and |V> – |H> is 45° to the left of vertical?"

"Exactly. They both give a fifty-fifty chance of being measured as horizontal or vertical, but they're different states. If you rotated your polarizer 45° clockwise, the |V> + |H> photons would all make it through, while the |V> – |H> photons would all be blocked."

*Strictly speaking, then, before its state is measured, Schrödinger's cat can be in one of two states: "alive plus dead," or "alive minus dead."

"So, State I is up and to the right, while State II is up and to the left?"

"Well, it'd be more complicated than that. There are two particles, so you'd need to do it in four dimensions, or something, but that's the basic idea."

"Okay, I guess I buy that. Wait—you said the original two entangled photons need to have opposite polarizations. Shouldn't they be in State III or IV, then?"

"You're absolutely right. In the usual teleportation procedure, Photons 2 and 3 need to be in State IV. I didn't mention that earlier, because I thought it would complicate things needlessly. Good catch."

"I'm a very smart dog. You can't get anything past me."

When Truman makes his measurement asking whether Photons 1 and 2 have the same polarization, Photons 1 and 2 are projected into one of these four states. At that instant, the entanglement between Photons 2 and 3 means that RD's Photon 3 is put into a definite polarization state that depends on which state Truman measured. There are four possible results for the polarization of RD's Photon 3, whose horizontal and vertical components are related to the horizontal and vertical components of Truman's original Photon 1.

Each result is a simple rotation of the original polarization state—the arrows point in a different direction, but still involve *a* steps in one direction (up, down, left, or right), and *b* steps in another. Given the outcome of Truman's measurement, RD knows how to recover the original state of Truman's photon, even though he doesn't know what that state was.

All Truman has to do, then, is call RD and tell him the result of the measurement. At that point, RD knows exactly what he needs to do to get Photon 3 into the right state. Based on the result of Truman's measurement, RD can rotate the polarization of Photon 3, and know that he's got exactly the state that Truman started with.

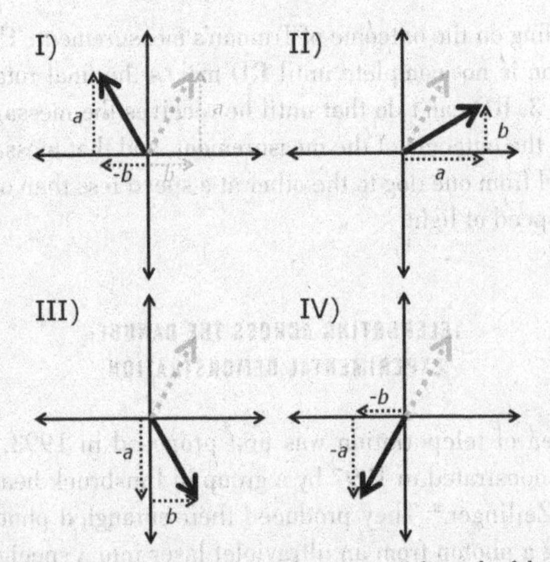

The state of RD's Photon 3 after "teleportation," for each of the four possible outcomes of Truman's entangling measurement. Each state is a simple rotation of the initial polarization of Photon 1 (dotted arrow).

This scheme transfers the polarization state of Photon 1 to Photon 3, transforming it into a perfect copy of the initial state of Photon 1. In the process, though, the entangling measurement made on Photons 1 and 2 has changed the state of Photon 1 so that it is no longer in the same state as when it started—it's in an indeterminate entangled state with Photon 2. It's impossible for both dogs to end up with exactly the same state, satisfying the no-cloning theorem.

We also see that teleportation is not instantaneous. The polarization of Photon 3 is instantaneously determined when Truman makes the measurement on Photons 1 and 2, but there's one more step, because Photon 3 is not instantaneously put into the *correct* state. Instead, it goes into one of four possible states, depending on the outcome of Truman's measurement. The teleportation is not complete until RD makes the final rotation of Photon 3. RD can't do that until he receives the message containing the outcome of the measurement, and that message has to travel from one dog to the other at a speed less than or equal to the speed of light.

TELEPORTING ACROSS THE DANUBE: EXPERIMENTAL DEMONSTRATION

The idea of teleportation was first proposed in 1993, and it was demonstrated in 1997 by a group in Innsbruck headed by Anton Zeilinger.* They produced their entangled photons by sending a photon from an ultraviolet laser into a special crystal that produces two infrared photons, each having half the energy of the original photon. They sent the laser through the crystal twice, to produce a total of four photons. One pair was used as the entangled pair needed for teleportation (Photons 2 and 3), while one of the other two was sent through a polarizer to provide the state to be teleported (Photon 1). The fourth photon was used as a trigger to let the experimenters know when to collect data.

Photons 1 and 2 were brought together on a beam splitter in a way that performed the entangling measurement. They could only detect one of the four Bell states, but when they did, they knew that Photon 3 was projected into a particular polarization. When they detected Photons 1 and 2 in State IV (25% of the time), they sent a signal to their analyzer to measure the polar-

*The same Anton Zeilinger was seen in chapter 1 heading the group that demonstrated diffraction of molecules, and chapter 5 doing quantum interrogation. He has made a long and distinguished career doing experiments to demonstrate the weird and wonderful features of quantum mechanics.

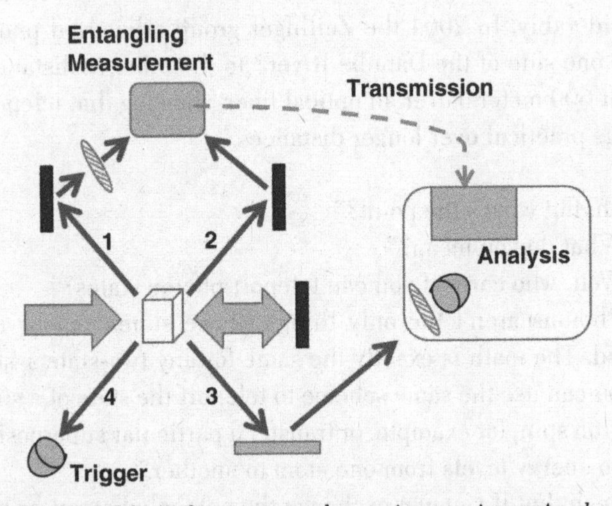

Schematic of the Zeilinger group teleportation experiment. An ultraviolet laser passes through a downconversion crystal, where it produces two infrared photons (Photons 2 and 3), which serve as the entangled pair for teleportation. The ultraviolet laser then hits a mirror, and passes through the crystal again, producing another pair (Photons 1 and 4), one of which serves as the photon to be teleported, while the other is a trigger to let the experimenters know that all four photons have been produced. Photons 1 and 2 are brought together for an entangling measurement, and when they are found in the appropriate Bell state, the polarization of Photon 3 is measured to confirm the "teleportation."

ization of Photon 3. Because they set the polarization of Photon 1 themselves, they were able to repeat the experiment many times, and confirm that Photon 3 was polarized at exactly the angle predicted by the teleportation protocol.

The initial demonstration used only one of the four possible Bell states in the measurement step, for reasons of experimental convenience, and teleported the polarization state all of half a meter. Subsequent experiments have expanded the measurements to include all four outcomes and extended the distance considerably. In 2004 the Zeilinger group teleported photons from one side of the Danube River* to the other (a distance of about 600 meters) over an optical fiber, showing that teleportation is practical over longer distances.†

"Yeah, but what's the point?"

"What do you mean?"

"Well, who cares if you can teleport photon states?"

"Photons aren't the only things whose states can be teleported. The math is exactly the same for any two-state system, so you can use the same scheme to teleport the state of a single electron spin, for example, or transfer a particular superposition of two energy levels from one atom to another."

"Yeah, but if you can exchange the entangled atoms or electrons, why don't you just send them, instead of teleporting them?"

"Atomic states and electron spins are kind of fragile, and it's hard to send them long distances without the state getting messed up. What you can do is to take an atom, say, and entangle it with one photon from an entangled pair, and use the other photon to teleport the state of the first atom onto another atom in a distant place."

"Okay, that's a little better, but it's still just one atom."

"It doesn't have to be. In 2006, a group at the Niels Bohr Institute in Copenhagen used teleportation to transfer a collective state from one group of atoms to another. There were about

*In the seven years between the two experiments, Professor Zeilinger moved from Innsbruck to Vienna.

†There's no inherent problem with sending photons over very long distances—light manages to reach us from distant galaxies, after all—but interactions with the environment can destroy entangled states through the process of decoherence discussed in chapter 4. The Vienna experiment shows that decoherence can be avoided long enough to send entangled photons over useful distances.

a trillion atoms in each of the two groups, which is still pretty small compared to dogs and people, but it shows that you can apply the technique to a larger system."

"That still sounds pretty useless, but I guess it's getting better."

"Thank you. You're very kind."

WHAT IS IT ALL FOR? APPLICATIONS OF TELEPORTATION

The quantum teleportation protocol lets us use entanglement to faithfully move a particular quantum state from one location to another, without physically moving the initial object. It can be used to reproduce photon states at distant locations, or to transfer a superposition state from one atom or group of atoms to another. Of course, it's still a long way from the science fiction ideal.

As with the classical fax machine, the only thing transmitted is information. Quantum teleportation allows us to transfer a particular state or superposition of states from one place to another, in the same way that the fax machine allows us to send a facsimile of what's printed on a paper document over telephone lines. If the state being "teleported" is the state of an atom, however, there have to be appropriate atoms waiting at the other end of the teleportation scheme, in the same way that the receiving fax machine needs to be loaded with paper and ink.

If the goal is to transfer an object from one place to another, though, it's not obvious that you *need* quantum teleportation. Quantum teleportation moves a particular state from one place to another, but if you're sending an inanimate object like a dog treat from one place to another, you may not need to preserve the exact state. As long as you have the right molecules in the right places relative to one another, it doesn't make much difference to the taste or texture of the treat if the atoms in the facsimile treat are not in precisely the same states as the original. All you really need is a fax machine that works at the molecular level, and there's nothing inherently quantum about that.

So why should we care about quantum teleportation? Quantum teleportation may not be needed to move inanimate objects, but it may be crucial for moving conscious entities. Some scientists believe that consciousness is essentially a quantum phenomenon—Roger Penrose, for example, promotes this idea in *The Emperor's New Mind*. If they're right, we would need a quantum teleporter, not just a fax machine, to move people or dogs, in order to properly reproduce their brain state. Quantum teleportation may be the key to ensuring that when Scotty beams you up to the *Enterprise*, you arrive thinking the same thoughts as when you left.

We're not even close to teleporting people, though, so the current interest in quantum teleportation involves much smaller objects. Quantum teleportation is useful and important for situations where state information is the critical item that needs to be moved from one place to another. The primary application for this sort of thing today is in quantum computing.

A **quantum computer**, like the classical computers we use today, is essentially a large collection of objects that can take on two states, called "0" and "1."* You can string these "bits" together to represent numbers. For example, the number "229" would be represented by eight bits in the pattern "11100101."

In a quantum computer, however, the "qubits"† can be found

*A classical computer uses millions of tiny transistors on silicon chips; quantum computers could use anything with at least two states—atoms, molecules, electrons.

†The "qu" is for "quantum." Physicists are not widely admired for their ability to think up clever names.

not just in the "0" and "1" states, but in superpositions of "0" and "1" at the same time. They can also be in entangled states, with the state of one qubit depending on the state of another qubit in a different location. These extra elements let a quantum computer solve certain kinds of problems much faster than any classical computer—factoring large numbers, for example. The modern cryptography schemes used to encode messages—whether they're government secrets or credit card transactions on the Internet—rely on factoring being a slow process. A working quantum computer might be able to crack these codes quickly, leading to intense interest in quantum computing from governments and banks.*

The precise quantum state of an individual qubit is critical to the functioning of a quantum computer, and it's here that quantum teleportation may find useful applications. A calculation involving a large number of qubits may require the entanglement of two qubits that are separated by a significant distance in the computer. Teleportation might be useful as a way of doing the necessary operations.

Further down the road, if we want to connect together two or more quantum computers in different locations, to make what Jeff Kimble of Caltech calls the "Quantum Internet," schemes based on entanglement and teleportation may be essential. This would allow still greater improvements in computing, in the same way that the classical Internet does for everyday computers.†

Whatever its eventual applications, quantum teleportation

*In fact, the National Security Agency is one of the largest funders of quantum computing research in the United States.

†Of course, it may also lead to quantum e-mail from dogs in Nigeria offering nine billion pounds of kibble if we'll just provide the bank account information to help with a simple transaction . . .

is a fascinating topic. It shows us that the nonlocal effects of quantum entanglement and the "spooky action at a distance" explored in the EPR paper can be put to use, moving information around in a way that can't be done by more traditional means. It may not help dogs to catch squirrels (not yet, anyway), but it's another source of insight regarding the deep and bizarre quantum nature of the universe.

"I don't know, dude. I still think it's lame."

"How's that?"

"Well, I mean, if you call something 'teleportation,' I expect it to be good for more than just moving state information."

"That is kind of unfortunate, I agree. I'm not the one who made up the name, though."

"So, that's it for entanglement, then? Just Aspect experiments and teleportation?"

"No, not at all. There are lots of things you can use quantum entanglement for. It's the key to quantum computing, as I said, and you can use it for 'dense coding,' sending two bits of information for every one bit transmitted."

"That's still just moving information around."

"There's also quantum cryptography, where you use entanglement to transmit a string of random numbers from one person to another, numbers that they can then use to encode messages in a completely secure way. There's no possibility of anyone eavesdropping on their messages, because the eavesdropping would change the particle states, and mess up the code in a way that can be detected."

"Still just information."

"Well, okay, sure, but there are people who think that the proper way to think about quantum physics is in terms of information. In some sense, the whole science of physics is really all about information."

"Yeah? Well, I'm a dog, and I'm all about getting squirrels."

"Okay, but that's really about information, too."

"How so?"

"Well, for your information, there's a big fat squirrel sitting right in the middle of the lawn."

"Ooooh! Fat, squeaky squirrels!"

CHAPTER 9

Bunnies Made of Cheese: Virtual Particles and Quantum Electrodynamics

Emmy is standing at the window, wagging her tail excitedly. I look outside, and the backyard is empty. "What are you looking at?" I ask.

"Bunnies made of cheese!" she says. I look again, and the yard is still empty.

"There are no bunnies out there," I say, "and there are certainly not any bunnies made of cheese. The backyard is empty."

"But particles are created out of empty space all the time, right?"

"You're still reading my quantum physics books?"

"It's boring here when you're not home. Anyway, answer the question."

"Well, yes, in a sense. They're called **virtual particles**, and under the right conditions, the zero-point energy of the vacuum can occasionally manifest as pairs of particles, one normal matter and one antimatter."

"See?" she says, wagging her tail harder. "Bunnies made of cheese!"

"I'm not sure how that helps you," I say. "Virtual particles have to annihilate one another in a very short time, in order to satisfy the energy-time uncertainty principle. A virtual electron-positron pair lasts something like 10^{-21} seconds before it disappears. They're not around long enough to be real particles."

"But they can become real, right?" She looks a little concerned. "I mean, what about **Hawking radiation**?"

"Well, sure, in a sense. The idea is that a virtual pair created near a black hole can have one of its members sucked into the black hole, at which point the other particle zips off and becomes real."

The tail-wagging picks back up. "Bunnies made of cheese!"

"What?"

She gives an exasperated sigh. "Look, virtual particles are created all the time, right? Including in our backyard?"

"Yes, that's right."

"Including bunnies, yes?"

"Well, technically, it would have to be a bunny-antibunny pair . . ."

"And these bunnies, they could be made of cheese."

"It's not very likely, but I suppose in a Max Tegmark* sort of 'everything possible *must* exist' kind of universe, then yes, there's a possibility that a bunny-antibunny pair made of cheese (and anticheese) might be created in the backyard, but—"

"And if I eat one, the other becomes real." She's wagging her tail so hard that her whole rear end is shaking.

"Yeah, but they wouldn't last very long before they annihilated each other."

"I'm very fast."

"Given the mass of a bunny, they'd only last 10^{-52} seconds. If that."

*Max Tegmark is a cosmologist at MIT known for proposing that our universe is one of a vast number of universes in a larger "multiverse." According to Tegmark, this multiverse contains every possible kind of universe that can be described mathematically, even those that would make no sense to us. Tegmark's work is somewhat similar to the "modal realism" of the philosopher David Kellogg Lewis.

"In that case, you'd better let me outside. So I can catch the bunnies made of cheese."

I sigh. "If you wanted to go outside, why didn't you just say that?"

"What fun would that be? Anyway, bunnies made of cheese!"

I look out the window again. "I still don't see any bunnies, but there is a squirrel by the bird feeder."

"Ooooo! Squirrels!" I open the door, and she goes charging outside after the squirrel, who makes it up a tree just in time.

Back in chapter 2, we saw that the wave nature of matter gives rise to zero-point energy, meaning that no quantum particle can ever be completely at rest, but will always have at least some energy. Incredibly, this idea applies even to empty space. In quantum physics, even a perfect vacuum is a constant storm of activity, with "virtual particles" popping into existence for a fleeting moment, thanks to zero-point energy, then disappearing again.

The idea of "virtual particles" popping in and out of existence in the middle of empty space is one of the most compelling and bizarre ideas in modern physics. In this chapter, we'll talk about quantum electrodynamics ("QED" for short), the underlying theory that gives rise to the idea of virtual particles. We'll also talk about the experiments that make QED arguably the most precisely tested theory in the history of science. Ironically, though, our discussion of this ultraprecise theory needs to start with the Heisenberg uncertainty principle.

COUNTING TAKES TIME: ENERGY-TIME UNCERTAINTY

The best-known version of the uncertainty principle is the one that we talked about in chapter 2 (page 48), which puts a limit on the uncertainties in the position and momentum of a particle. At a very fundamental level, the more we know about the position of a particle, the less we can know about how fast it is moving, and vice versa.

Slightly less well known is the uncertainty relationship between energy and time, which says that the uncertainty in energy multiplied by the uncertainty in time has to be larger than Planck's constant divided by four pi:

$$\Delta E\ \Delta t \geq h/4\pi$$

As with position-momentum uncertainty, this means that the more we know about one of these two quantities, the less we can know about the other.

The idea that energy is somehow related to time may seem strange at first, but we can understand it by thinking about light. As we saw in chapter 1 (page 21), the energy of a photon is determined by the frequency associated with that color of light. A low-uncertainty measurement of energy, then, requires a precise measurement of frequency.

So, how do you make a precise measurement of frequency? Imagine that you want to measure the rate at which an excited dog is wagging her tail, which is a fairly regular oscillation with small fluctuations in the frequency and amplitude: sometimes she wags a little faster, sometimes a little slower, sometimes farther to the right, sometimes farther to the left. What is the best way to measure the wagging frequency?

Frequency is measured in oscillations per second, so you need to count the number of wags that take place in some fixed time interval. If you wait five seconds, and count ten tail wags, that's a frequency of two oscillations per second. Any such measurement will always have some uncertainty, though—when you counted ten oscillations, was it really ten full oscillations, or ten-and-a-little-bit? Had her tail gotten all the way to the right, or was she wagging it farther that time around?

To minimize that uncertainty, you need to look over a much longer time. The uncertainty in your count of wags will tend to be a constant—one tenth of a tail wag, say—so the more oscillations you count, the better you do in terms of the relative uncertainty.

If you watch a tail wagging for five seconds, and count ten oscillations, plus or minus one tenth, the frequency you measure is

$$f = (10 \pm 0.1 \text{ oscillations})/(5 \text{ seconds}) = 2.00 \pm 0.02 \text{ Hz*}$$

That is, the frequency is somewhere between 1.98 and 2.02 oscillations per second.

If you watch for fifty seconds (assuming the dog doesn't explode from impatience), you'll count a hundred wags, plus or minus one tenth, and the frequency is then

$$f = (100 \pm 0.1 \text{ oscillations})/(50 \text{ seconds}) = 2.000 \pm 0.002 \text{ Hz}$$

Increasing the number of oscillations that you measure leads to a decrease in the uncertainty of the frequency, and a more precise determination of just how happy the dog is.

The cost of that decrease in frequency uncertainty is an increase in the time uncertainty. To get one tenth the energy uncertainty, you spent ten times as long making the measurement, which means you can't say exactly *when* you measured the frequency. You know the *average* frequency over those fifty seconds, but you can't point to a specific instant and say that the frequency right then was 2.000 Hz. All you can say is that over that fifty-second interval, the dog's tail was wagging at

*The units for frequency are "Hertz," abbreviated Hz, after the German physicist Heinrich Hertz, who was the first physicist to demonstrate experimentally that light is an electromagnetic wave.

about 2 Hz, but at any given instant, it may have been faster or slower.

You could be more specific about the measurement time, by measuring for only half a second, say, and counting one tail wag, but then the frequency uncertainty becomes much larger: $f = 2.0 \pm 0.2$ Hz. You can't have a small uncertainty in both the frequency of an oscillation and the time when it was measured.

The same logic applies to measuring the frequency of light, though the oscillations are much too fast to count by hand. We directly see this uncertainty principle in action in the interaction between light and atoms. We know from chapters 2 and 3 that atoms will be found only in certain allowed energy states, and that they move between these states by absorbing or emitting photons.

When an atom moves from a high-energy state to a lower-energy state, the frequency of the emitted photon is determined by the energy difference between the two states. That energy difference has some uncertainty, though, that depends on how much time the atom spent in the higher-energy state. Two identical atoms placed in the same high-energy state can thus emit photons with very slightly different frequencies.

The difference is tiny—for typical atoms, it's about one hundred-millionth of the frequency of the photons. This can be measured using lasers, though, and this tiny frequency uncertainty limits our ability to make certain measurements of atomic properties.

"So it takes you awhile to count things. Big deal. What's that have to do with bunnies made of cheese?"

"The frequency-counting example is just one example of a more general principle. **Energy-time uncertainty** holds no matter what form the energy is in."

"Why?"

"Well, all forms of energy have to be equivalent, because

you can convert one form of energy to another. So, if you have a photon of uncertain energy, and you use it to start an electron moving, the kinetic energy of that electron must be uncertain."

"I still don't see what this has to do with bunnies."

"We know from Einstein's theory of relativity that mass and energy are equivalent—"

"$E = mc^2$!"

"Exactly. Since mass is just another form of energy, you can convert energy into mass and mass into energy. Mass has uncertainty just like all the other forms of energy, and that uncertainty is related to how long the mass stuck around."

"So a bunny made of cheese would have an uncertain mass?"

"Right. If it was around for only a short time, the uncertainty could be very large—for something like a top quark, which sticks around for only 10^{-25} seconds, the quantum uncertainty due to that lifetime is close to 1% of the total mass."

"But if it was around for a long time, the mass uncertainty would be small? I want a bunny made of cheese with a small mass uncertainty!"

"Good luck with that. A bunny made of cheese would have an awfully short lifetime around you."

"Ooh. Good point."

WHEN THE HUMANS ARE AWAY . . . : VIRTUAL PARTICLES

How does this get us bunnies made of cheese? Well, let's think about applying this uncertainty principle to empty space. If we look at some small region over a long period of time, we can be quite confident that it is empty. Over a short interval, though, we can't say for certain that it *isn't* empty. The space might contain some particles, and in quantum mechanics, that means it will.

Uncertainty about the emptiness of space isn't as strange as it may seem at first. If a physicist or a stage magician gives a dog a box to inspect at leisure, she can conclusively state that the box is empty. She can sniff in all the corners, check for false bottoms, and make absolutely sure that there's nothing hiding in some little recess. If she's allowed only a brief peek or a quick sniff inside the box, though, she can't be as confident that the box is empty. There might be something tucked into a corner that she wasn't able to detect in that short time.

The amount of time needed to determine whether the box is empty also depends on the size of the thing you might expect to find. You don't need to look for very long to determine whether the box contains Professor Schrödinger's famous cat, but if you're attempting to rule out the presence of a much smaller object—a crumb of a dog treat, say—a more thorough inspection is required, and that takes time.*

The same idea applies to empty space in quantum physics, via the energy-time uncertainty relationship. When we look at an empty box over a long period of time, we can measure its energy content with a small uncertainty, and know that there is only zero-point energy—no particles are in the box. If we look over only a short interval, however, the uncertainty in the energy can be quite large. Since energy is equivalent to mass through Einstein's famous $E = mc^2$, this means that we can't be certain that the box *doesn't* contain any particles. And as with Schrödinger's cat, if we don't know the exact state of what's in the box, it's in a superposition of all the allowed states. The cat is both alive and dead, and the box is both empty and full of all manner of particles, *at the same time.*

To put it another way, the box can contain some particles, as long as they appear and disappear fast enough that we don't see them directly. But how can we arrange for particles to disappear so conveniently? Ordinary particles don't disappear when we're

*It's even harder for humans, whose noses don't work well enough to sniff out the really tiny crumbs in the corners.

not looking, unless they're edible and there's a dog in the area.

Particles can disappear from the box provided they come in pairs, one matter and one **antimatter**. Every particle in the universe has an antimatter equivalent with the same mass and the opposite charge—the antiparticle for an electron is a positron, the antiparticle for a proton is an antiproton, and so on. When a particle of ordinary matter comes into contact with its antiparticle, the two annihilate each other, converting their mass into energy.

In practical terms, this means that a particle and its antiparticle can pop into existence from nothing inside the box. This temporarily increases the energy inside the box by a small amount—an electron-positron pair increases the energy by two electron masses multiplied by the speed of light squared—but as long as they mutually annihilate in a short enough time, there's no problem, because the uncertainty in the energy is large enough to cover two extra particles.*

How long do we need to look to rule out any possibility of the box containing a particular type of particle? Well, we need the uncertainty in the energy to be smaller than the energy that the particle has from Einstein's $E = mc^2$, so the time will depend on the mass. To make the energy uncertainty small enough to rule out the possibility of the box containing one electron and one positron, we need only to look at the box for 10^{-21} seconds. That's a billionth of a trillionth of a second—slightly less than the time required for light to travel from one side of an atom to the other. A proton and antiproton have a greater mass, and would last even less time—10^{-24} seconds or less. A bunny with

*Or, to put it yet another way, if the particle-antiparticle pair is around for only a short time, their masses have a large uncertainty. If the time is short enough, the uncertainty in the mass can be larger than the mass, in which case we can't say that the mass *wasn't* zero. In which case, they can exist for that short time, because two particles with zero mass don't increase the energy inside the box at all.

a mass of a kilogram (whether it's made of cheese or something else), would last for 10^{-52} seconds,* or 0.0000000000000000 0000000000000000000000000000000001 s.

"I think you're making this much more complicated than it needs to be. How hard is it to measure zero particles?"

"This, from a dog who has to sniff the *entire* kitchen floor every night, on the off chance that we dropped a molecule of food while making dinner?"

"But you *do* drop food sometimes, and I like your food better than mine."

"The point is, it's really difficult to measure zero. You can never really say that something has a definite value of zero, only that the value is no bigger than the uncertainty associated with your measurement. Some people spend their whole scientific careers measuring things to be zero, with better and better precision."

"That sounds pretty depressing."

"It does require a certain personality type. Anyway, that's kind of a side issue, because we know that virtual particles exist. We can detect their effects."

"Wait, if these things only stick around for such a short time, how do you detect them?"

"We can't detect them, not directly. We know they exist because we can see the effects they have on other particles. Virtual particles popping in and out of existence change the way that real particles interact with one another."

"How does that work?"

"Well . . ."

*This is an interval so short it may literally have no meaning—in some exotic theories, time itself is quantized, and comes in steps of 10^{-44} s (the "Planck time"). In such models, it's impossible to have an interval shorter than that.

EVERY PICTURE TELLS A STORY: FEYNMAN DIAGRAMS AND QED

Virtual particles, particles that appear and disappear too quickly to be seen directly, may seem too fanciful for a serious scientific theory. In fact, they are absolutely critical to the theory known as **quantum electrodynamics** or QED, which describes the interaction between light and matter at the most fundamental level. QED describes all interactions between electrons, or between electrons and electric or magnetic fields, in terms of electrons absorbing and emitting photons.

The best-known form of QED relies on pictures called **Feynman diagrams**, after the noted physicist and colorful character Richard Feynman, who invented them as a sort of calculational shortcut. These diagrams represent complex calculations in the form of pictures that tell a story about what happens as particles interact.* The simplest Feynman diagram for an electron interacting with an electric or magnetic field looks like this:

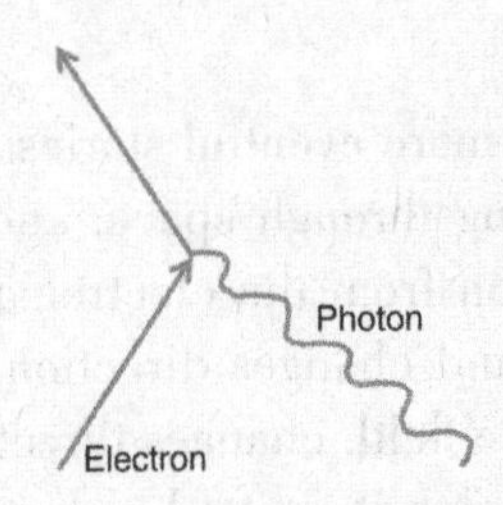

*Feynman was famous for his very intuitive approach to physics, and the diagram-based technique he developed gained wide acceptance because it provides a convenient way of thinking about the complex calculations required for QED. At the same time Feynman was doing his work, Julian Schwinger developed a much more formal approach to the same problems, giving the same results as the Feynman approach in a more mathematically rigorous way. Both approaches are widely used in theoretical physics today, and Feynman and Schwinger shared the 1965 Nobel Prize with Shin-Ichiro Tomonaga, who independently developed some of the same techniques as Schwinger.

The straight lines represent an electron moving through some region of space, and the squiggly line represents a photon from the electromagnetic field. In this diagram, time flows from the bottom of the picture to the top of the picture, while the horizontal direction indicates motion through space, so the diagram itself is a story in pictures: "Once upon a time, there was an electron, which interacted with a photon, and changed its direction of motion."

You might think that this little story, boring as it is, is the complete picture of what happens when an electron interacts with light, but it turns out there are myriad other possibilities, thanks to virtual particles. For example, we can have diagrams like these:

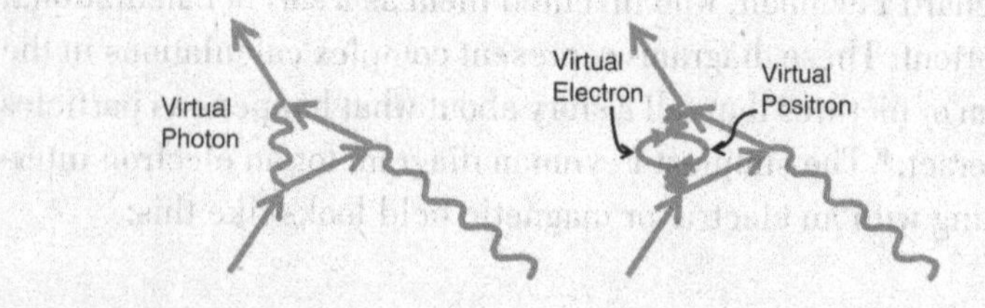

These tell much more eventful stories: on the left, our electron is moving along through space, and just before it interacts with the photon from the electromagnetic field, it emits a virtual photon* and changes direction. Then it absorbs the real photon from the field, changes direction again, then reabsorbs the virtual photon it emitted earlier. The right diagram is even stranger: the virtual photon spontaneously turns into an electron-positron pair,† which then mutually annihilate, turning back into a photon, which gets reabsorbed.

*In a Feynman diagram, any particle that appears and disappears before the end of the "story" is a "virtual particle."

†A positron can be mathematically described as an electron moving backward in time, another trick invented by Feynman, hence the downward arrow.

These diagrams show the odd relationship between virtual particles and the normal laws of physics. The virtual particles seem to disobey some basic rules of physics—a real electron could never move fast enough to get back in front of a photon to reabsorb it—but virtual particles can break the rules as long as they're not around long enough to violate energy-time uncertainty. It's similar to the relationship between a dog and furniture: when humans are looking, being on the couch is strictly forbidden, but as long as there are no humans around, and she gets off before they see her, it's a great place to nap.

Each of these Feynman diagrams represents a tiny slice of the story of what might happen to an electron interacting with a photon from an electric or magnetic field. Each of these diagrams also stands for a calculation in QED giving the final energy of the electron, and the likelihood of the events depicted. As you increase the number of virtual particles, the diagrams become much less likely, but they remain possible, as long as they happen very quickly.

The diagrams on the previous page only involve electrons, positrons, and photons, but any type of particle can turn up as a virtual particle. The likelihood of a particle appearing decreases as its mass increases, so a virtual proton-antiproton pair is much less likely than a virtual electron-positron pair (a proton has almost 2,000 times the mass of an electron), but in principle, there is no limit. If you wait long enough, you should expect any and every type of object showing up as a virtual particle—even bunnies made of cheese.

"Okay, so virtual particles affect the way electrons interact with photons. Big deal. Why should I care?"

"The photons in those Feynman diagrams could represent any sort of electric or magnetic interaction. Those diagrams could be describing an electron interacting with another electron, or with a proton, and that happens all the time."

"Okay, I have no idea what you're talking about."

"Well, when you're dealing with QED, you talk about interactions in terms of the exchange of particles. Two electrons that repel each other do so by passing a photon from one to the other—one emits a photon, which travels over to the other, and gets absorbed. The absorption and emission change the momentum, and we see that as a force pushing the two particles apart."

"That sounds complicated. Why would you think about it that way?"

"It turns out to be more convenient, mathematically. You can also look at it as a natural way to incorporate the fact that nothing can travel faster than light. In the classical picture of electric forces developed before Einstein, when you change the position of one electron, the force on the other should change instantaneously, no matter how far away it is. That goes against relativity, which says that there's no way to transmit anything faster than light."

"Oh. That's a problem."

"That's right. If we think of the forces as arising from photons which are passed from one particle to another, though, that takes care of the problem. The photons travel at the speed of light, and the force doesn't change until the photons get there."

"So, the photons in those diagrams . . ."

"They could be real photons, due to the electron interacting with a magnet or an electric field, or they could be photons emitted by another electron. Either way, the effect is the same—the presence of virtual particles changes the interaction, and we can detect that."

"You still haven't explained how to detect it, though."

"I'm getting there. If you'd stop interrupting . . ."

THE MOST PRECISELY TESTED THEORY IN HISTORY: EXPERIMENTAL VERIFICATION OF QED

Virtual particles come and go in an extremely short time, faster than we can directly observe. We know they exist, though, because interactions with virtual particles change the way electrons interact. The effect is tiny, but we can measure it, and it agrees with experimental observations to fourteen decimal places. Not only do these experiments confirm that QED is correct, they may provide a way to detect the existence of new subatomic particles *without* using billion-dollar particle accelerators.

How does this work? Well, because the virtual particles are around for such a short time, we don't know which of the many possibilities took place for any given electron. Quantum mechanics tells us that if we don't know the exact state of a particle, it exists in a combination of all the possible states—a superposition state like those we've discussed in previous chapters. So, when physicists calculate the effect of an electron interacting with an electric or magnetic field, they need to include all the possible Feynman diagrams describing the process.*

In a certain sense, as the electron passes from point A to point B, it follows all the possible paths from A to B at the same time. It absorbs a single real photon, but also emits and reabsorbs a virtual photon, and that virtual photon does and doesn't turn into an electron-positron pair, and so on. All of these processes are possible, so they all contribute to the superposition of Feynman diagrams.

*Of course, there are an infinite number of things that *might* happen, and thus an infinite number of possible diagrams. In practice, though, the more complicated the diagram, the smaller the contribution it makes to the answer, so theoretical physicists need to add up only as many diagrams as they need to match the precision of an experiment.

Another way of thinking about this is to note that we never see a real electron interacting with a single photon; rather, what we see is the cumulative effect of a great many repeated interactions. If we could watch each of these in detail, we would see that most of the time, the electron simply absorbs the photon, with no virtual funny business. Maybe one time in ten thousand, it will emit a virtual photon. One virtual photon in ten thousand will create a virtual electron-positron pair, and so on. Each time you get one of these different processes, it changes the total energy by a small amount, and that needs to be taken into account.

In this view, the electron is like a dog walking down the street. The chances of the dog stopping to sniff any given plant are very small,* but there are dozens of plants along the side of any suburban street, and one of them is bound to smell fascinating, and demand further investigation. The human holding the leash will need to take that delay into account when planning the walk.

We don't have the ability to watch each of the interactions of a real electron in sufficient detail to keep track of just how many times virtual particles were involved in the interaction, any more than we can predict exactly which plants the dog will stop to sniff, but we can model the effect that this has on the interaction. Adding up Feynman diagrams can be thought of as calculating the *average* change in the energy of the electron from absorbing one real photon. We then describe the cumulative effect of many photon absorptions, which may or may not involve virtual particles, as a series of absorptions with the same average energy, in the same way that we can describe the progress of a dog by saying that he will stop three times a block on average. It may be five times on some blocks, and only one time on others, but over the entire walk, the effect is the same as if he stopped three times on every block.

Whether you think of it as a superposition of all possible

*Unless the dog is a basset hound, in which case he'll want to thoroughly sniff *everything*.

paths at the same time, or an average interaction being used to cover the details of many individual interactions, the effect of virtual particles shows up when we look at an electron interacting with a magnetic field. Electrons (and all other material particles) have a property called "spin" that makes them act like tiny little magnets, with north and south poles.* The energy of an electron whose poles are aligned with the magnetic field is very slightly different than the energy of an electron whose poles point in the opposite direction.

The energy difference between these states in a magnetic field depends on a number called the **gyromagnetic ratio,** or **g-factor,** of the electron, which basically tells you how big a magnet you get for a given amount of "spin." The simplest quantum theory of the electron says that the value of this ratio should be exactly 2, which is what you would get if there were no virtual particles. Thanks to the contributions of virtual particles, the actual value is very slightly higher.

This value can be measured with extraordinary precision. In 2008, a team of physicists led by Gerald Gabrielse of Harvard made the most accurate measurement to date of the electron g-factor in experiments involving single electrons held in a Penning trap. The experimental value they obtained was:

$$g = 2.00231930436146 \pm 0.00000000000056$$

They compared their result to the result of a QED calculation that involved summing almost a thousand individual Feynman diagrams†—including processes much more complex than

*The name "spin" is because this is similar to what you would see if the electron were a spinning ball of charge. The electrons are not literally spinning, but the math is the same.

†An impressive achievement in its own right, done by the group of Toichiro Kinoshita at Cornell.

those described above—and theory and experiment were in perfect agreement, to fourteen decimal places. This is why we can say with confidence that virtual particles exist, no matter how strange the idea may seem at first glance.

While this agreement is extremely impressive, a *disagreement* would be even more interesting. The g-factor for the electron includes only the effects of virtual photons and virtual electron-positron pairs, but the analogous quantity for the muon (a type of particle similar to an electron, but with a larger mass) shows the effects of a number of other, more exotic, virtual particles.* The latest measurements of the g-factor of the muon show a tiny difference between the experimental value and the theoretical prediction. This difference might be merely a calculation or measurement error, or it might hint at the existence of a new particle that wasn't included in the calculations—some particle that isn't part of the standard model of particle physics. It's still much too early to say for sure, but if this difference holds up, it could be the first experimental test of the many theories for physics beyond the standard model.

Experimentalists have a long way to go before they start to see the effects of virtual bunnies made of cheese, of course. But if it is possible to make bunnies out of cheese, we know that they must be out there somewhere.

"So that's what you mean when you say 'Quantum mechanics is the most accurately tested theory in the history of theories.' "

"Well, I didn't say it like *that* . . ."

"Whatever. You say it all the time, and it sounds really pompous."

*The muon g-factor calculations include virtual muons, tau particles, quarks, and gluons, which account for most of the known types of subatomic particles.

"Fine. See if I rub your belly anymore. Anyway, yes, that's what I mean. Quantum electrodynamics has been used to predict the g-factor of the electron to fourteen decimal places, and it agrees perfectly with experimental measurements. And QED is a relatively straightforward extension of ordinary quantum mechanics to situations where you need to think about relativity."

"So, what else is it good for?"

"Well, as I said, you can have all sorts of different things as virtual particles. This includes particles that have been predicted by theorists, but not observed yet—if they really exist, they'll show up as virtual particles."

"How does that help anything?"

"Some of those hypothetical particles would allow interactions that aren't possible with any of the particles we know about. In addition to the muon g-factor, there's the 'electric dipole moment' of the electron. If the right sort of particles exist, they would change the way an electron interacts with electric fields inside atoms and molecules. It might be possible to detect the tiny shift in the allowed states in experiments with lasers."

"So, if you saw this dipole moment thing, you would know that new types of particles exist?"

"And if you *don't* see it, you can rule out some types of particles. The fact that nobody has observed an electric dipole moment in an electron yet has ruled out some of the simpler models that theorists like to use. If some of the new experiments don't see anything, it could really make the theorists squirm."

"That's pretty cool."

"It's especially cool because these are tabletop experiments, not billion-dollar particle accelerators. There are a whole bunch of groups doing these experiments—at Berkeley, Yale, Washington, Colorado, and other places—and they're our best chance of learning something really new until the Large Hadron Collider starts up, and maybe beyond that."

"Of course, none of this helps with what's really important."

"That being?"

"Providing me with bunnies made of cheese."

"Well, what can I say? In some areas, physics still has a long way to go."

"I'll say."

CHAPTER 10

Beware of Evil Squirrels: Misuses of Quantum Physics

I'm putting seed in the squirrelproof bird feeder when I hear a little voice above my head. "Pssst! Hey, human dude!"

There's a squirrel perched on the branch of a tree, staring at me. I look around, but Emmy is on the far side of the yard, intently sniffing the base of the big oak tree. "What do you want?" I ask.

"How about you give me some of that seed?"

"I don't think—what's with your face? Is that a goatee?"

"It's a fake. We wear 'em to mess with the dog—she thinks we're from another dimension. Look, how about *selling* me some birdseed, then?"

"What's a squirrel going to use to buy birdseed?"

"How's free energy sound, hmm?" Somehow, despite the fake goatee, fuzzy tail, and protruding teeth, he manages to look smug.

"Free energy?"

"Yeah, we can tell you how to extract a nearly infinite amount of energy from ordinary water. That ought to be worth some birdseed, eh?"

"Really. Free energy."

"You bet. We extract the zero-point vibrational energy from water molecules, leaving them in a lower energy state than ordinary molecules. You can turn that energy directly into electricity, and use it to power lights, or computers, or birdseed-making machines." Looking closely, I see that the fake goatee is obviously held on by string.

"Sounds too good to be true. What's the catch? Toxic waste products?"

"No, no—the only waste is still water. In fact, it's better than water. It's superwater!"

"What's so super about it?"

"Well, it's in a different quantum state, right? So it's got, like, special properties and stuff. You can drink it, and it'll cure diseases."

"How's that work?"

"Well, you drink it, and concentrate on measuring your wavefunction to be in a healthy state. If you do it right, you can think your way to perfect health."

"You don't say."

"Yeah, we run some workshops and classes and stuff. I'm a hundred and six, and in perfect health. For just a handful of birdseed a week, we'll let you in on the secret."

"Uh-huh." The dog is still on the far side of the yard, carefully checking the bushes for bunnies.

"You can also use it to power a quantum computer, but that'll cost more than birdseed. For a jar of peanut butter a week, you can have the schematics of a quantum computer that you can use to crack the encryption on credit card transactions."

"Wow."

"I know. Pretty cool, huh?"

"The dog was right. You *are* evil squirrels."

"Yeah, like you wouldn't do it if you knew how. So, how 'bout that birdseed?"

"Sure, I'll give you some birdseed . . ." I put a little pile of seed down on the ground, about six feet from the trunk of the tree.

"Thanks, buddy," says the squirrel, scampering down the trunk. "You're a prince."

"Don't mention it," I say, stepping between the squirrel and the tree. "Emmy!" Across the yard, her head snaps around. "Look! Evil squirrel!"

"Ooooh!" She comes charging across the yard, teeth bared. The squirrel tries to run back up the tree, but I block his path, so he turns and flees toward the maples at the back of the yard, with the dog snapping at his tail.

I spot something in the grass, and bend down to retrieve a tiny fake goatee. I drop it in the trash on my way back into the house.

In the preceding chapters, we have talked about a lot of weird and wonderful features of quantum mechanics. Wave-particle duality, quantum measurement, EPR correlations, virtual particles—so many aspects of quantum theory defy our everyday experience that quantum mechanics starts to seem like magic. None of the normal rules seem to apply, and it may look as though absolutely anything is possible.

This is a common misconception regarding quantum mechanics, and you'll find it repeated in lots of places. A little time with Google will turn up dozens of sites offering "quantum" methods to produce energy for nothing, improve your health and well-being, or even amass wealth and power: lots of people out there are making money by peddling quantum mechanics as magic.

Quantum mechanics is *not* magic, though. No matter how unlikely or amazing it seems, quantum mechanics is a scientific theory that has to conform to the general principles of physics. The word "quantum" in the description of a phenomenon or device does not allow it to create energy out of nothing or send messages faster than the speed of light. These principles are built into the deep structure of the universe. Quantum mechanics is not only compatible with those rules, but in some cases, these rules arise from quantum behavior.

While many of the predictions of quantum mechanics seem to defy our everyday intuition for how the world works, they do not suspend all the rules of common sense. In particular, they do not supersede the most important commonsense rule for dealing with the world: if something sounds too good to be true, it almost certainly is.

We've spent the bulk of this book talking about the wonderful features of quantum theory, but I want to close on a cautionary note. There are a lot of people peddling a false version of quantum mechanics as magic, offering results beyond your most wildly unrealistic dreams. Some of them are scam artists, and some of them are sincere but deluded, but they're all wrong. The hucksters are hard to separate from the merely confused, but it is not difficult to spot false versions of quantum theory, and in this chapter, I'll point out a few of the most common problems.

A "QUANTUM" FREE LUNCH: FREE ENERGY

One of the two main areas worked by scam artists abusing quantum theory is the field of "free energy." Free energy scammers always claim to have developed a scheme that will produce huge amounts of energy for a trivial amount of work. This takes lots of different forms, but the basic appeal is always the same: you put a small amount of work in, and get a large amount of electricity out. All it will take is a small investment of cash to get the prototype system working, and soon you'll be out from under the thumb of the power company forever . . .

This is nothing more than a claim to have invented a perpetual motion machine, and scientists have known for hundreds of years that perpetual motion machines are impossible. Quantum mechanics does not change that conclusion.

The most common phony explanation for a "quantum" perpetual motion machine is that it is tapping the zero-point energy of some system or another. This zero-point energy is the energy that quantum physics tells us is present even in a system in its lowest possible energy state. When you've extracted all the energy that you can from a quantum system—reduced the kinetic energy to its lowest possible value and removed outside interactions that would raise the potential energy—there is still some residual energy left in the system.

Hucksters like to point to this zero-point energy as a resource to be tapped. "There's still energy there," they say, "and our device taps that energy to keep the perpetual motion machine moving."

We saw in chapter 2 (page 52), though, that zero-point energy exists because matter is fundamentally wavelike, and quantum particles must always have some wavelength. For the energy of a system to truly be zero, it would have to be perfectly still at a specific position, and that is impossible for any system described by a wave. Zero-point energy, like the Heisenberg uncertainty principle, is a consequence of the fundamental wave nature of matter. Just as there is no way to evade the limits imposed by uncertainty, there is no way to extract the zero-point energy to do useful work. Trying to use the zero-point energy is like asking for half a photon—it's a request that makes no sense.

Probably the most successful proponent of free energy via bogus quantum theory to date is a company called Black Light Power. Bob Park of the University of Maryland and the American Physical Society has spent nearly twenty years debunking the claims of the company's founder, Randell Mills, which is well covered in Park's book *Voodoo Science: The Path from Foolishness to Fraud* (Oxford, 2000). Despite Park's efforts, though, Black Light Power is still around, peddling a remarkable energy-generating process in which (according to their website) "energy is released as the electrons of hydrogen atoms are induced by a catalyst to transition to lower-energy levels (i.e., drop to lower base orbits around each atom's nucleus) corresponding to fractional quantum numbers." These mysterious lower-energy hydrogen atoms, called "hydrinos," are claimed to have all sorts of magical properties, supposedly enabling new high-voltage batteries and miraculous light sources (none of which are available yet, but they're promised any day now).

This sounds impressively science-y, but even a dog can tell that it's nonsense. Hydrogen is the simplest atom in the universe, consisting of a single proton orbited by a single electron. The first quantum model of hydrogen was put forward by Niels Bohr in 1913, and a full quantum treatment was developed using the Schrödinger equation in the 1920s. Quantum electrodynamics was first applied to hydrogen in 1947, and modern QED models of hydrogen agree with experimental results to the same phenomenal precision as the measurements of the electron g-factor. The hydrogen atom is one of the best understood and most precisely tested systems in the universe.

Modern physics leaves no room for states "below the ground state" in hydrogen. For such states to exist, our understanding of fundamental physics would need to be so far wrong that it would be impossible to achieve the fourteen-decimal-place agreement between experiment and theory that we see with QED.

Another commonly cited source of "quantum" free energy is the "vacuum energy." This is just a variant of the zero-point energy scheme that purports to tap the zero-point energy of empty space—the constantly appearing and disappearing sea of virtual particles that QED shows must exist.

"Vacuum energy" schemes are no more possible than "hydrino" power. Empty space does contain energy in the form of virtual particles, but those particles appear at random, and disappear again in a tiny fraction of a second. We have no way of making electrons appear on demand, nor do we have any way of making them stick around to do useful work. Vacuum energy exerts a small but real influence on electrons and other particles, but it is not an energy resource that can be tapped.

Any claim of perpetual motion or "free energy" is essentially a claim that you can get something for nothing. Any dog knows, deep down, that that's not possible—you don't get treats without doing some sort of a trick. Throwing the word "quantum" around doesn't change that basic fact: you can't get something from nothing. Anyone who claims otherwise is selling something.

"I thought you said virtual particles did become real, as in Hawking radiation?"

"Sort of. The idea is that an electron and a positron can appear right at the edge of a black hole, in such a way that one of them falls into the black hole, while the other escapes."

"Right, so electrons are created out of nothing!"

"Wrong. When a virtual electron becomes real through the Hawking process, the black hole actually *loses* a bit of mass, to make up for it. The energy to make the real electron doesn't come from the vacuum energy, it comes from the black hole, which gets whittled away to nothing, one virtual particle at a time."

"So, I guess it's not much use as a power source, then?"

"No, not really. Even assuming you could contain and control a black hole, it would be consumed through the power generation process, just like anything else. There's no such thing as a free lunch."

"Yes there is. You never charge me for lunch. Or breakfast, or dinner, or snacks . . ."

"You earn your food by being cute."

"Oh, yeah. And protecting the house from evil squirrels!"

"That, too. Speaking of which . . ."

MEASURING YOUR WAY TO HEALTH: "QUANTUM HEALING"

The other main source of misused quantum mechanics is "alternative" medicine. Bookstores and the Internet abound with people pitching quantum mechanics as the key to health, wealth, and long life.

The most common form of these claims involves quantum measurement. Hucksters note that in quantum theory states aren't determined until they are measured. They then claim that the key to health is simply to measure yourself as healthy. You can live forever, in this line of thinking, by running a quantum Zeno effect experiment on yourself—if you're always measuring yourself to be in fine health, quantum measurement will see to it that you never get sick.

The best example of this line of quantum quackery is Deepak Chopra, who even has a book titled *Quantum Healing* (Bantam, 1990), the first of a string of bestselling alternative medicine books. What is "quantum healing," you ask? Chopra offers a one-paragraph explanation in a 1995 interview:

> Quantum healing is healing the bodymind from a quantum level. That means from a level which is not manifest at a sensory level. Our bodies ultimately are fields of information, intelligence and energy. Quantum healing involves a shift in the fields of energy information, so as to bring about a correction in an idea that has gone wrong. So quantum healing involves healing one mode of consciousness, mind, to bring about changes in another mode of consciousness, body.*

He uses a lot of scientific-sounding terms, but this is just word salad. It has all the scientific validity of the technobabble on old episodes of *Star Trek*—all he's missing is a call to "reverse the polarity" of something.

He expands on these ideas in *Ageless Body, Timeless Mind* (Harmony, 1994) whose subtitle promises a "Quantum Alterna-

*The interview is online at http://www.healthy.net/scr/interview.asp?ID=167, and was retrieved in summer 2008.

tive to Getting Old." His explanation of the physical basis of his ideas shows an impressive ignorance of history, boldly declaring that "Einstein taught us that the physical body, like all material objects, is an illusion, and trying to manipulate it can be like grasping the shadow and missing the substance" (p. 10). This is almost exactly the opposite of Einstein's view, as we saw in chapter 7. Einstein was profoundly disturbed by the idea of quantum indeterminacy, which Chopra takes to an entirely new level, arguing that nothing actually exists:

> Because there are no absolute quantities in the material world, it is false to say that there even is an independent world "out there." The world is a reflection of the sensory apparatus that registers it . . . All that is really "out there" is raw, unformed data waiting to be interpreted by you, the perceiver. You take "a radically ambiguous flowing quantum soup" as physicists call it,* and use your senses to congeal the soup into the solid three-dimensional world. [p. 11]

While he acknowledges that this descent into solipsism may sound "disturbing," perhaps because congealed soup sounds unappealing, he sees this as a feature, not a bug, writing that "there is incredible liberation in realizing that you can change your world—including your body—*simply by changing your perception*." (pp. 11–12) In other words, since nothing really exists, you might as well be healthy, wealthy, and youthful. It's all just a matter of "perception," which is to say, measurement.

The idea of "quantum healing," using the active nature of quantum measurement to ensure good health, suffers from two

*The plural "physicists" here is probably not justified—"radically ambiguous flowing quantum soup" is not a common phrase in physics. The only actual physicist who has ever used it appears to be Nick Herbert, a promoter of "Quantum Tantra," which is about what you would expect.

major problems. The first problem is that Chopra and other authors are applying quantum ideas to systems that are far too large to show quantum effects. As we've seen again and again throughout this book, quantum effects are extremely difficult to tease out, and the larger the system being studied, the harder it is to see quantum effects. The largest object ever seen in a quantum superposition state is a collection of about a billion electrons (see chapter 4, page 101), while the quantum Zeno effect (chapter 5) has only been demonstrated with single particles.

The bigger problem, though, is that quantum measurements are fundamentally random. The state of a quantum system is indeterminate until a measurement is made, and the specific outcome of an individual measurement cannot be predicted. It doesn't matter whether you subscribe to a Copenhagen-like interpretation in which the wavefunction collapses to a single value, or a many-worlds-like interpretation in which you simply perceive a single branch of an ever-expanding wavefunction, or even Chopra's congealed soup interpretation: there is no way to know in advance how a given quantum measurement will turn out.

There may very well be a branch of the wavefunction of the universe in which every dog enjoys perfect health and an infinite supply of fat, slow bunnies, but there is no way to influence measurement outcomes to reach that universe. Meditation doesn't help, positive thinking won't get the job done, drugs don't do any good—there is no known way to influence the quantum structure of the universe to generate a particular outcome of a quantum measurement, and no scientific study has ever detected a hint of one. If it were possible to achieve great things simply by wanting them badly enough, physicists would have a much easier time demonstrating quantum effects,* and dogs would never lack for steak, cheese, and bunnies.

*Not to mention securing funding for their experiments.

Meditation may lower your stress level, and thinking positive thoughts may improve your mood, and either of those things may make you feel better about your lot in life and thus help you find the energy to catch that bunny. There's nothing quantum about that, though, and you're not tapping into the deep structure of the universe in any meaningful way.

"You're right about the meditation thing. Meditating really helps me lower my stress level."

"Since when do you meditate?"

"Since always. I have Buddha nature. I like to meditate in the sun in the backyard."

"That's not meditating, that's sleeping. Your eyes are closed, and you snore."

"That's not snoring, that's . . . a mantra."

"You're ridiculous. What stress do you have, anyway?"

"Oh, my life is very hard. I have to worry about whether to sleep in the living room, or the dining room, or the office. I worry about why you're not petting me, why you're not giving me treats . . ."

"Okay, stop it. You're giving me a headache."

"You should try meditating!"

SPOOKY HEALING THROUGH ENTANGLEMENT: "DISTANT HEALING"

Another common form of quantum quackery is claiming quantum nonlocality as a basis for "alternative" or "traditional" medicine. The claim is that the correlations seen in Bell's theorem and the Aspect experiments show that there is some deeper, "transcendent," level of reality. This supposedly produces a connection between all living things, and practitioners can use this connection to diagnose problems or even heal people without actually touching them. It's also the alleged basis for all sorts of ESP phenomena.

The most extreme variant of this idea is found in books like *Distant Healing*, by Jack Angelo (Sounds True, 2008), which proclaims:

> Quantum mechanics scientists believe that the unified field theory connects everything in the universe including gravity, nuclear reactions, electromagnetic fields, and human consciousness. Modern physics, then, supports the finding of Distant Healing that thought forms, such as ideas and information, are able to travel from one part of the human family to another via a network of consciousness. [pp. 180–81]

Again, this is *Star Trek*–level stuff. There is no "unified field theory" in physics—indeed, the *lack* of a unified field theory is one of the great challenges of modern physics. Even if there were a unified field theory, "human consciousness" is not one of the elements it would include.

The idea is fleshed out in more detail, with more erroneous justifications, in Tiffany Snow's *Forward from the Mind: Distant Healing, Bilocation, Medical Intuition & Prayer in a Quantum World* (Spirit Journey Books, 2006):

> *Entanglement* (sometimes called *nonlocality*) is where a *faster-than-the-speed-of-light* signal is instantaneously communicated between two particles which somehow remain in touch with each other no matter how far apart they are. What happens to one instantly happens to the other, even across a galaxy, through the entangled waveform connection. And if we look at the beginning of the universe through a "Big Bang" theory, we see all energy (which includes us) was entangled in the very beginning, so we all have an indent [*sic*] with each other now, even though we may be far apart. Very simply, what one of us does, does *affect all others*. [p. 31]

This starts off with an explanation that is only slightly wrong, but it goes completely off the rails by the end of the paragraph.

As we saw in chapter 7, entanglement does, indeed, allow for nonlocal correlations between the states of entangled particles. However, these correlations must first be established through local interactions—the photons in the Aspect experiments were initially produced by the same atom, for example.

The quantum connection between these entangled particles is extremely fragile, and it's easily broken by interactions with the rest of the universe, leading to decoherence. Physicists have to work very hard to produce an entangled state that lasts even a tenth of a second. No entanglement-based connection could possibly survive the fourteen billion years since the Big Bang.

Since there is no residual entanglement from the Big Bang, there is no inherent connection between separated objects. I can easily arrange for the states of two dogs to become correlated, as discussed in chapter 7 (page 143), but only by bringing those dogs together and allowing them to interact with each other. If the two dogs are always separated, there is no way for them to become entangled. Similarly, even leaving aside the fact that human organs are far too large to show quantum effects, there is simply no way to establish a connection between, say, the liver of a patient and the hands of a "healer," without some contact between them.

Entanglement is also invoked as an explanation for homeopathy. In homeopathic treatments, minute quantities of herbs or toxins are placed in water and then diluted to a point where there should not be a single molecule of the original herb or toxin in a given water sample. Homeopaths claim that the water "remembers" the presence of the original substance, though, and acquires some of its properties, which supposedly enables the water to heal patients who drink it. Entanglement is sometimes cited as an explanation for this "memory" effect, with the claim being that the interaction between the water and the herb or toxin establishes a connection along the lines of the correlations seen in the Aspect experiments.

The absolute pinnacle of the quantum-entanglement explanation of homeopathy is presented in the work of Lionel Milgrom, who theorizes that it involves not merely entanglement between water and toxin, but a three-way entanglement between the patient, the practitioner, and the remedy (he calls this "PPR" entanglement). Milgrom has a wonderful ability to ape the jargon and notation of quantum physics, and writes in a 2006 paper:

> [I]t should be possible to use notions of quantum entanglement (and by implication, information processing) to illustrate certain features of the therapeutic process in homeopathy and other CAMs.* Consequently, the effects of investigating homeopathy and other CAMs using blinded trial procedures should also be amenable to such illustration. Thus, in double-blinded provings, each of the components in the PPR entangled state may be thought of as two-state "macro-qubits" . . . and, therefore, by implication, the homeopathic process might be considered to involve macro-quantum "teleportation." . . . However, it is only the entangled state which contains information about the whole system. Thus, anything which breaks the entangled state will necessarily lead to loss of information about the integration of function of the systems as a whole system. Clearly, this could happen in [double-blind randomized control trials†] of homeo-

*"CAM" = "complementary and alternative medicine." Acronyms make anything sound more scientific.

†Double-blind randomized control trials are medical tests in which patients are randomly selected to receive either the treatment being tested or a placebo, and neither the patient nor the doctor dispensing the treatment knows which is which. These are the gold standard for modern medical research.

pathic efficacy, where either the remedy or patient and practitioner are removed from their entangled therapeutic context.*

This is a truly breathtaking application of "quantum" reasoning. Not only does Milgrom attribute the curative effects of homeopathic remedies to this "macro-quantum 'teleportation,'" but in a lovely bit of logical judo, he uses this argument to explain away the failure of homeopathic remedies to outperform placebos in properly run clinical trials. It's all quantum, you see, and thus attempting to measure the performance in a manner consistent with scientific principles ruins everything. Milgrom concludes that standard medical testing protocols simply can't be used to measure homeopathy because "they seem to destroy the very effects they were purportedly designed to investigate," an impressive attempt to use quantum mechanics to avoid (possibly via tunneling or teleportation) meeting the exacting standards applied to conventional medical treatments.

The claim that entanglement explains homeopathy is patent nonsense. Patients and practitioners are much too large to exhibit quantum behavior, and even though there is the slight possibility of an entangling interaction between molecules in a solution, entanglement is just a correlation between the *states* of two systems—when an atom is in a particular state, a photon is vertically polarized, or when one dog is awake, another dog is also awake. This does not mean that one system has acquired characteristics of the other. A quantum interaction can set up a correlation between an atom and a photon, but it can't turn an atom into a photon or water into healing elixir, any more than an interaction between two dogs can turn a Labrador retriever into a Boston terrier.

*Lionel R. Milgrom, *Evidence-Based Complementary and Alternative Medicine* 4, 7–16 (2006). Quotes from p. 14.

BEWARE OF EVIL SQUIRRELS: QUANTUM PHYSICS IS NOT MAGIC

Quantum mechanics is a weird and wonderful theory, and it makes some amazing things possible. Most modern technology depends on quantum mechanics in one way or another—modern electronic devices and computer chips rely on quantum effects in order to operate, and optical devices like the lasers and LEDs used in modern telecommunications are fundamentally quantum devices. Quantum theory may also provide for future technologies with amazing potential—quantum computers that can solve problems faster than any classical computer, or quantum cryptography systems that protect messages using unbreakable codes.

As astounding as its results are, though, quantum mechanics does not provide a basis for miracles. Its predictions defy everyday intuition, but the theory does not completely override common sense. If somebody promises results that sound too good to be true, chances are they're lying, either to you or to themselves. Dropping a few quantum buzzwords into the explanation doesn't make free energy or eternal youth any more plausible.

There are many fascinating aspects of quantum theory that we haven't talked about here. There are also a lot of evil squirrels in the world, with or without goatees. Think carefully about claims made regarding quantum effects, and keep in mind that while it may be weird, quantum mechanics is not magic. If you do that, you'll have no trouble finding the wonderful aspects of our quantum universe, and avoiding quacks and crackpots.

"Wow, dude, that's pretty depressing."

"What? Quantum theory doesn't need to be magic to be cool."

"No, not that. I'm talking about the scammers. I knew squirrels were evil, but I didn't know there were humans who were that bad."

"Yeah, it's a little depressing to see a good theory abused in this way. But on some level, it's a sign of progress."

"How do you figure?"

"Well, if you go back to the 1800s, you can find people making the same sorts of magical claims about electricity. All sorts of nonsensical devices were proposed that were supposed to do magical things because they used electricity."

"Yeah?"

"And in the mid-1900s, it was atomic or nuclear power. People suggested using nuclear power for the most absurd things, and any number of scams claimed atomic power as a basis."

"Yeah? What's your point?"

"Well, nobody really falls for either of those anymore. We've gotten used to electricity and nuclear power, and people no longer believe ridiculous claims made about them."

"So 'quantum' is the new 'atomic'?"

"Pretty much. Scammers have had to move on to 'quantum' as an explanation, because the old explanations don't work anymore. So, in a sense, the fact that people are peddling 'quantum' hokum means that the general public has gotten a tiny bit less credulous over the years. 'Quantum' still works because most people don't know what it means."

"So you need to teach more people about quantum."

"Exactly. Hence this book."

"And I'm helping! I'm a public-service dog!"

"You're a very good dog."

"So, is that it for the book, then?"

"Pretty much. Why?"

"Well, if you're done with the book, can we go for a walk?"

"Sure."

"And if we see any evil squirrels . . ."

"If we see evil squirrels, you can bite them."

"Ooooh!"

Acknowledgments

I learned about the physics described in this book from a large number of mentors and colleagues over a period of almost twenty years. Many thanks are due to Bill Phillips, Steve Rolston, Paul Lett, Kris Helmerson, Ivan Deutsch, Aephraim Steinberg, Luis Orozco, Paul Kwiat, Mark Kasevich, Dave DeMille, Seyffie Maleki, Kevin Jones, Jeff Strait, Stuart Crampton, and Bill Wootters. All the good parts of the explanations are ultimately due to them; any mistakes are original to me.

I got many helpful comments on an early draft of this book from my intrepid beta readers: Jane Acheson, Lisa Bao, Aaron Bergman, Sean Carroll, Yoon Ha Lee, Matt McIrvin, and Frances Moffet. Michael Nielsen and David Kaiser also made helpful comments on draft copies. All of them helped make this a better book than it would've been otherwise.

This book grew out of a couple of posts on my weblog, "Uncertain Principles" (http://scienceblogs.com/principles/), which eventually became the opening dialogues of chapters 4 and 9. Thanks are due to the folks at ScienceBlogs—Christopher Mims, Katherine Sharpe, Erin Johnson, and Arikia Millikan—for providing me with a platform, and to Cory Doctorow of Boing Boing and the people at Digg for promoting those posts. Barrett Garese, Erin Hosier, and Patrick Nielsen Hayden deserve thanks for convincing me that writing a physics book with my dog was a good idea. And of course, thanks to my editor, Beth

Wareham, and agent, Erin Hosier, for all their help getting the book into shape, and helping me navigate the publishing process, which is completely different than anything in physics.

Emmy was adopted from the Mohawk & Hudson River Humane Society shelter in Menands, New York (http://www.mohawkhumanesociety.org/). Like most animal shelters, they are an excellent source of wonderful dogs (and other pets), and I would encourage anyone thinking of getting a dog to look at their local shelter.

I've been lucky enough to know a lot of dogs over the years—Patches, Rory, Truman, the late great RD, Bodie, and even Tinker—and there's a little bit of all of them in this book. Most of the credit goes to Emmy, though, who is unquestionably the best Emmy ever, and the Queen of Niskayuna.

Many thanks are due to my friends and family, who have been tremendously supportive despite finding the whole thing a little weird. And last, but far from least, thanks to my wife, Kate Nepveu, for reading innumerable drafts and gently correcting my grammar; for patiently listening to me rant and kick ideas around; and for baby Claire, who complicated things in the best way possible, and most of all for inspiring the whole thing by laughing when I have silly conversations with the dog. This quite literally would not have happened without her.

Further Reading

David Lindley's *Uncertainty: Einstein, Heisenberg, Bohr, and the Struggle for the Soul of Science* (Doubleday, 2007) provides a very readable introduction to the early history of quantum theory, as well as a detailed account of the debates about the theory and the meaning of the uncertainty principle.

Louisa Gilder's *The Age of Entanglement: When Quantum Physics Was Reborn* (Knopf, 2008) covers some of the same territory as *Uncertainty*, but with more of an emphasis on entanglement, and goes on to describe Bohm's nonlocal hidden variable theory, Bell's theorem, and the first experimental tests of nonlocality. The book is built around several reconstructed conversations among the important figures in the story, with dialogue pieced together from letters and memoirs.

The Tests of Time: Readings in the Development of Physical Theory, edited by Lisa M. Dolling, Arthur F. Gianelli, and Glenn N. Statile (Princeton University Press, 2003), reproduces many of the classic papers in early quantum theory, including Bohr's original model of hydrogen; the Einstein, Podolsky, and Rosen paper; Bohr's response to EPR; and John Bell's famous theorem.

The Theory of Almost Everything: The Standard Model, the Unsung Triumph of Modern Physics by Robert Oerter (Plume, 2006) gives an excellent overview of the state of modern physics, including basic quantum theory, QED, and issues in particle physics that are not discussed in this book.

Richard Feynman's *QED: The Strange Theory of Light and Matter* (Princeton University Press, 2006) is a wonderfully accessible explanation of the details of quantum electrodynamics. His autobiographical books (*Surely You're Joking, Mr. Feynman* and *What Do You Care What Other People Think?*) have less physics, but are great fun.

For more mathematically inclined readers, *The Quantum Challenge, Second Edition: Modern Research on the Foundations of Quantum Mechanics* by George Greenstein and Arthur G. Zajonc (Jones and Bartlett, 2005) gives an excellent overview of many of the experiments that have demonstrated the strange features of quantum mechanics.

James Gleick's *Genius: The Life and Science of Richard Feynman* includes a description of the development of QED, and the rivalry between Feynman and Julian Schwinger.

On an artistic note, Michael Frayn's play *Copenhagen* makes powerful use of quantum ideas in exploring the famous falling-out between Niels Bohr and Werner Heisenberg.

And finally, George Gamow's "Mr. Tompkins" stories (collected in *Mr. Tompkins in Paperback* [Cambridge University Press, Canto Imprint, 1993]) are the original whimsical exploration of modern physics, through the daydreams of an unassuming bank clerk. No physics book involving a talking dog could possibly fail to mention them.

Glossary of Important Terms

allowed state: One of a limited number of states in which an object may be measured in quantum mechanics. For example, a dog at rest can either be found on the floor, or on the couch, but never halfway between the floor and the couch.

antimatter: Every particle in the universe has an antimatter equivalent, with the same mass and the opposite charge. When a particle of ordinary matter encounters its antiparticle, the two annihilate, converting their mass into energy.

Bell's theorem: A mathematical theorem proved by John Bell, showing that entangled quantum particles have their states correlated in ways that no local hidden variable (LHV) theory can match.

classical physics: Physics theories developed before about 1900, describing the behavior of everyday objects. Core components are Newton's laws of motion, Maxwell's equations for electricity and magnetism, and the laws of thermodynamics.

coherence: A property of waves or wavefunctions, roughly defined as behaving as if the waves came from a single source. Adding together two coherent waves gives a clear interference pattern; adding together two incoherent waves gives a rapidly shifting pattern that smears out and becomes indistinct. The process of "decoherence" destroys the coherence between two waves from a single source through random and fluctuating interactions with a larger environment.

conservation of energy: The law of conservation of energy states that energy can be changed from one form to another, but the total energy of a given system is always the same. For example, a dog can convert the potential energy stored in food into kinetic energy as she chases a squirrel, but she cannot gain more kinetic energy than the total energy available in her food.

Copenhagen interpretation: The philosophical framework for quantum mechanics developed by Neils Bohr and colleagues at his institute in Denmark. The Copenhagen interpretation insists on an absolute split between microscopic systems, which are described by quantum mechanics, and macroscopic systems, which are described by classical physics. The interaction between a microscopic quantum system and a macroscopic measuring apparatus causes the wavefunction to "collapse" into one of the allowed states for that system.

decoherence: A process by which random, fluctuating interactions with the environment destroy our ability to see an interference pattern for a quantum particle. Decoherence is particularly important for the many-worlds interpretation, where it ensures that different branches of the wavefunction of the universe will not affect one another.

diffraction: A characteristic behavior of waves, in which waves passing through a narrow opening or around an obstacle spread out on the far side. A dog can hear a potato chip hitting the kitchen floor from the living room because sound waves diffract through the kitchen door and around corners.

Einstein, Podolsky, and Rosen (EPR) paradox: A famous paper by Albert Einstein, Boris Podolsky, and Nathan Rosen that used entangled particles to argue that quantum mechanics was incomplete. Their argument has been proven wrong by experiments testing Bell's theorem, but has led to the development of quantum teleportation and other quantum information technologies.

energy: A measure of an object's ability to change its own motion or the motion of another object. Energy comes in many forms, such as kinetic energy, potential energy, and mass energy (Einstein's $E = mc^2$). Energy may be converted from one form to another, but cannot be created or destroyed.

energy-time uncertainty: A variation of the Heisenberg uncertainty principle stating that it is impossible to know both the exact energy of some object and the exact time at which it was measured. This limits the lifetime of virtual particles in quantum electrodynamics.

entanglement: A quantum "connection" between two objects whose states are correlated in such a way that measuring one also determines the state of the other. A classical analogy is two dogs in the same room: either both will be awake, or both will be asleep. If you measure one dog to be awake, you immediately know that the other is also awake. Similar correlations exist for quantum particles, but their states are indeterminate until one of the two is measured, at which time the state of both is instantaneously determined, no matter how far apart they are.

Feynman diagram: A picture representing a possible sequence of events for a charged particle interacting with light. Each diagram stands for a calculation in QED, and the energy of the interacting particle is found by adding together all the possible diagrams for that particle. The diagrams are named after Richard Feynman, who invented them as a calculational shortcut.

gyromagnetic ratio/ "g-factor": A number, given the symbol *g*, that determines how an electron interacts with a magnetic field. The simplest theory of quantum mechanics predicts that $g = 2$, but QED predicts a value that is very slightly larger. The experimentally measured value of *g* agrees with the QED prediction to fourteen decimal places.

Hawking radiation: A process by which "virtual particles" cause black holes to evaporate. When a particle-antiparticle pair appears near a black hole, one of the two can fall into the black hole, while the other escapes. In order to conserve energy, the black hole must lose a tiny bit of mass. Over time, the black hole is whittled down to nothing, one particle mass at a time.

interference: A phenomenon that occurs when two or more waves are added together. If the peaks of one wave line up with the peaks of the other ("in phase"), the result is a much larger wave. If the peaks of one wave line up with the valleys of the other ("out of phase"), the result is no wave at all. Interference patterns involving single particles are the clearest demonstration of quantum behavior.

kinetic energy: Energy associated with a moving object. For everyday objects, the kinetic energy is equal to half the mass times the speed squared ($\frac{1}{2} mv^2$). A Great Dane has more kinetic energy than a Chihuahua moving at the same speed, while a hyperactive Siberian husky has more kinetic energy than a sleepy bloodhound of the same mass.

local hidden variable (LHV) theory: A theory of the sort preferred by Einstein, Podolsky, and Rosen. In an LHV theory, measurements made in one position are independent of measurements made at other positions ("local"), and particles are always in definite states, though the exact values are unknown ("hidden variables"). LHV theories cannot duplicate all the predictions of quantum mechanics (according to Bell's theorem), and have been disproven in experiments by Alain Aspect, among others.

many-worlds interpretation: The philosophical framework for quantum mechanics developed by Hugh Everett III at Princeton in the 1950s. The many-worlds interpretation avoids the "wavefunction collapse" problem of the Copenhagen interpretation by saying that all possible measurement outcomes take place in different branches of the wavefunction—in some part of the wavefunction, every dog eats steak. Sadly, we only perceive a single branch. The other branches of the wavefunction are effectively separate universes, due to decoherence, which prevents the different branches from having a measurable effect on one another.

measurement: In quantum mechanics, an active process that changes the state of the system being measured. Before a measurement is made, a quantum object will be in a superposition of all the allowed states; after the measurement, the object will be in one and only one state. The Copenhagen interpretation and the many-worlds interpretation offer two different ways of describing what happens during a measurement.

modern physics: Physics theories developed after about 1900, consisting principally of relativity and quantum mechanics.

momentum: A quantity associated with motion that determines what will happen to an object during a collision. In classical physics, momentum is mass times velocity ($p = mv$); a small Chihuahua must be moving much faster than a Great Dane to have the same momentum. In quantum mechanics, the momentum of a particle determines its wavelength, through the de Broglie relation $\lambda = h/p$.

no-cloning theorem: A mathematical theorem showing that it is impossible to make a perfect copy of a quantum object without knowing its state in advance.

particle-wave duality: A feature of quantum mechanics, in which objects have both particle and wave properties. Classical physics says that light is a wave, but quantum physics says that a beam of light is also a stream of photons. Classical physics says that an electron is a particle, but quantum physics says that an electron also has a wavelength that depends on its momentum. Electrons and photons are each "quantum particles," a third class of object that is neither particle nor wave, but has properties of each.

photoelectric effect: An effect discovered in the late 1800s, where light falling on metal knocks out electrons. Einstein explained the photoelectric effect in 1905 by applying Planck's quantum hypothesis to light directly, describing a beam of light as a stream of photons.

photon: A "particle" of light. A beam of light may be thought of as a stream of particles, like kibble being poured into a dog's bowl. Each photon has an energy given by Planck's constant multiplied by the frequency associated with that color of light ($E = hf$).

Planck's Constant (h): The constant that relates energy to frequency or momentum to wavelength in quantum physics. The measured value is $h=6.6261\times10^{-34}$ J-s, or 0.0000000000000 000000000000000000006261 J-s, which is a very small number indeed.

polarization: A property of light, corresponding to the direction of oscillation of the light wave in classical physics. Any polarization can be described as a combination of vertical and horizontal polarizations, and will determine the probability of that light passing through a vertical or horizontal polarizer.

polarizer/ polarizing filter: A material that allows through light polarized at some angle, and blocks light polarized 90° away from that angle. Light polarized at some intermediate angle has a probability of passing through that depends on the angle.

potential energy: Energy associated with an object that is not currently moving, but has the potential to start moving. A dog always has potential energy, even when sleeping: at the slightest sound, she can leap up and start barking at nothing.

probability: The wavefunction for a quantum object describes the probability of finding the object in any of its allowed states when a measurement is made. For example, there is a high probability of finding the dog in the kitchen, a high probability of finding the dog in the living room, and a very low probability of finding the dog on the couch, if she knows what's good for her.

quantum computer: A computer made up of "qubits" that not only can take values of "0" and "1" like the bits in a classical computer, but also superpositions of "0" and "1." Such a computer could solve certain types of problems, such as the factoring of large numbers, much faster than a classical computer. The difficulty of factoring large numbers is the basis for modern cryptography, so a quantum computer would let an evil squirrel decipher your credit card transactions, and clean out your bank account to buy birdseed.

quantum electrodynamics: "QED" for short. The theory describing the interactions between charged particles and light, developed by Richard Feynman, Julian Schwinger, and Shin-Ichiro Tomonaga around 1950. Feynman's formulation is the best known version, which describes interactions in terms of the exchange of "virtual particles."

quantum eraser: A demonstration of quantum measurement in which an interference pattern is destroyed by making it possible to measure the exact path taken by a particle, but recovered by doing something to confuse that measurement.

quantum field theory: A theory that combines quantum mechanics with Einstein's relativity, to cover particles moving at speeds close to the speed of light, and the interactions between such particles. The simplest quantum field theory is quantum electrodynamics.

quantum interrogation: A technique using the quantum Zeno effect to detect the presence of an object without letting it absorb even a single photon. Its applications to bunny stalking should be obvious to any dog.

quantum mechanics/ quantum physics/ quantum theory: The subject of this book, quantum mechanics was developed in the first half of the twentieth century, and describes the behavior of and interactions among atoms, molecules, subatomic particles, and light.

quantum teleportation: A procedure for transferring the exact state of a quantum particle from one place to another without measuring it or moving it, using entangled particles as a resource. Sadly, it does not allow dogs to beam themselves into places where they can easily catch squirrels.

quantum Zeno effect: A demonstration of quantum measurement in which an object can be prevented from changing states by repeatedly measuring its state. The classical equivalent is a dog who prevents her owner from napping by constantly asking "Are you asleep?"

relativity: The theory developed by Albert Einstein to describe gravity and the behavior of objects moving at speeds close to the speed of light.

Schrödinger equation: The mathematical formula that physicists use to find the wavefunction for a particular quantum system, and predict how it changes in time.

Schrödinger's cat: A thought experiment proposed by Erwin Schrödinger, intended to show the absurdity of quantum superpositions. He imagined a cat enclosed in a box with a device that had a 50% chance of killing the cat within one hour; quantum physics says that at the end of the hour the cat is equal parts alive and dead, until its state is measured. This experiment has made him a hero to canine physicists.

semiclassical argument: A description of a physical system that is mostly based on classical physics, with a few modern ideas added in an ad hoc manner. Examples of semiclassical models include the "Heisenberg microscope" (page 38) and the Bohr model of hydrogen (page 49).

state/ quantum state: A particular collection of properties (position, momentum, energy, etc.) describing an object. For example, "sleeping in the living room," "sleeping in the kitchen," and "running around the house" are three different possible states for a dog.

superposition state: In quantum mechanics, an object can exist in a superposition of two or more allowed states at the same time, until a measurement is made. Such superposition states give rise to interference patterns, which can be detected experimentally, even though the system can only be measured in one allowed state.

thermal radiation: Also called "black-body radiation," the light that is emitted by a hot object, such as the characteristic red glow of a hot burner on a stove. The spectrum of this light depends only on the temperature of the object. Explaining this spectrum led Max Planck to introduce quantum mechanics.

tunneling: A quantum phenomenon in which a particle that does not have enough energy to pass over a barrier passes through the barrier anyway, appearing on the other side, like a bad dog digging a hole under a fence.

uncertainty principle/ Heisenberg uncertainty principle: One of a set of mathematical relationships limiting the precision with which complementary properties can be measured. The best known uncertainty principle is between momentum and position, and says that it is impossible to know both exactly where a bunny is, and exactly how fast it is moving. Any attempt to specify the position more precisely will lead to increased momentum uncertainty, and vice versa. The energy-time uncertainty relationship is also important, as it determines the length of time that virtual particles can exist in QED.

virtual particle: A particle in a Feynman diagram that appears and disappears too quickly to be measured directly. These usually appear as pairs of one normal particle and one antimatter particle, most often one electron and one positron. In principle, anything can show up as a virtual particle, even a bunny made of cheese.

wavefunction: A mathematical function whose square gives the probability of finding an object in any of its allowed states. In quantum mechanics, all objects are described by wavefunctions.

zero-point energy: The tiny amount of energy that is always present in a quantum object, thanks to the wave nature of matter. Confined quantum particles are never perfectly at rest—they're like puppies in a basket, always squirming and wiggling and shifting around, even when they're asleep.

CPSIA information can be obtained
at www.ICGtesting.com
Printed in the USA
LVHW100201290522
719808LV00002B/2